Materials for Nanophotonics — Plasmonics, Metamaterials and Light Localization

MATERIALS RESEARCH SOCIETY
SYMPOSIUM PROCEEDINGS VOLUME 1182

Materials for Nanophotonics — Plasmonics, Metamaterials and Light Localization

Symposium held April 14–17, 2009, San Francisco, California, U.S.A.

EDITORS:

Luca Dal Negro
Boston University
Boston, Massachusetts, U.S.A.

Mark Brongersma
Stanford University
Stanford, California, U.S.A.

Joseph M. Fukumoto
Northrop Grumman Space Technology
Redondo Beach, California, U.S.A.

Lukas Novotny
University of Rochester
Rochester, New York, U.S.A.

Materials Research Society
Warrendale, Pennsylvania

Single article reprints from this publication are available through
University Microfilms Inc., 300 North Zeeb Road, Ann Arbor,
Michigan 48106

CODEN: MRSPDH

Published by:

Materials Research Society
506 Keystone Drive
Warrendale, PA 15086
Telephone (724) 779-3003
Fax (724) 779-8313
Web site: http://www.mrs.org/

Manufactured in the United States of America

METAMATERIALS

OPTICAL NANOANTENNAS AND DECAY ENGINEERING

APPENDIX
This Appendix contains papers from
Materials Research Society Symposiun Proceedings Volume **1077E**,
Functional Plasmonics and Nanophotonics, S. Maier and S. Kawata, Editors

*Invited Paper

PREFACE

This proceeding volume collects the papers presented at Symposium EE, "Materials for Nanophotonics — Plasmonics, Metamaterials and Light Localization," held April 14–17 at the 2009 MRS Spring Meeting in San Francisco, California. The symposium covered fundamental aspects and the advanced applications of nanoplasmonics, and it was intended to bridge, for the first time, three closely related research fields: nanoplasmonics, metamaterials, and light localization in complex media.

The engineering of light localization, optical dispersion and plasmonic fields in complex optical media has the potential to boost the scaling of optical technologies below the diffraction limit, opening unprecedented opportunities for basic and applied research.

This volume gathers peer-reviewed papers that address key challenges in materials fabrication, nanoplasmonics, and metamaterials.

The Editors would like to thank all the authors of the proceedings papers and the reviewers who, by responding promptly and by providing their thoughtful comments, have improved the quality of this special issue and allowed us to present a contribution that is both timely and of high-scientific quality. Finally, we greatly thank Prof. Lukas Novotny and Dr. Joseph M. Fukumoto for their significant contributions across all the stages of the symposium organization.

Luca Dal Negro
Mark Brongersma
Joseph M. Fukumoto
Lukas Novotny

August 2009

MATERIALS RESEARCH SOCIETY SYMPOSIUM PROCEEDINGS

MATERIALS RESEARCH SOCIETY SYMPOSIUM PROCEEDINGS

Prior Materials Research Society Symposium Proceedings available by contacting Materials Research Society

Poster Session I

Mater. Res. Soc. Symp. Proc. Vol. 1182 © 2009 Materials Research Society   1182-EE09-08

Real-Time Visualization Method of Surface Plasmon Resonance With Spectroscopic Attenuated Total Reflection

Kensuke Murai[1], Masahiro Tsukamoto[2], Koji Furuichi[3, 4], Akitaka Kashihara[3, 5], Yasuo Takigawa[3]
[1]Photonics Research Institute, National Institute of Advanced Industrial Science and Technology (AIST), 1-8-31 Midorigaoka, Ikeda, Osaka 563-8577, Japan
[2]Joining and Welding Research Institute, Osaka University, 11-1 Mihogaoka, Ibaraki, Osaka 567-0047, Japan
[3]Department of Electronic Engineering and Computer Science, Osaka Electro-Communication University, 18-8 Hatsu-cho, Neyagawa, Osaka 572-8530, Japan
[4]Present Address: Research and Development Center, Dainippon Screen MFG. Co., Ltd., Tenjinkita-cho 1-1, Teranouchi-agaru 4-chome, Horikawa-dori, Kamigyo-ku, Kyoto 602-8585, Japan
[5]Present Address: Department of Development and Engineering, Shinko Seiki Co., Ltd., Miyake-cho 30, Moriyama, Shiga 524-0051, Japan

ABSTRACT

We report the real-time visualization technique of surface plasmon resonance (SPR) with the spectroscopic attenuated total reflection (S-ATR). Using this technique, light absorption due to SPR on thin metal films by the white light were clearly visualized on a two-dimensional detector such as CCD camera in the coordination of the incident angle and the wavelength. Metal films could be distinguished even in partially oxidized condition. The real-time SPR visualization would be useful for monitoring of biochemical reactions as well as for fabricating plasmonic devices.

1. INTRODUCTION

Recently, surface plasmons resonance (SPR) is widely studied for plasmonics devices for constructing faster processor in the electronic microprocessors as well as for biochemical evaluation. SPR is a specific absorption which is sensitive to the dielectric constant and thickness of the metallic material, resulting in strong absorption of light as a function of incident angle, wavelength and polarization of light.[1-15] Since the dispersion property of SPR exists in the total reflection region, it had been studied with the attenuated total reflection (ATR) method. Kretchmann configuration[3] using a glass prism and an approximately 50 nm-thick silver or gold film was often used as sensing instruments in order to evaluate the optical constants of the film itself,[3-7] the coating layer on the film,[5,8,9] the surface process deferences,[10,11] or the ambient material.[12] Most of SPR signals had been measured with either angular or spectral dependence with this geometry. In the case of angular dependence,[3-11,13] the monochromatic laser, e.g. He-Ne laser at 632.8 nm, is often used for the incident light. One can measure reflection loss as a function of an incident angle in the total reflection region. In the spectral dependence,[12] a white light source is used instead of a monochromatic laser in order to measure reflection loss as a function of wavelength at a fixed incident angle. In both cases, the absorption peak shifts easily due to a few nm-thick layer on the metal film.[1] Visualization technique of SPR on the thin film had been developed using a two-dimensional detector such as a CCD camera.[17-20]

On the other hand, another imaging technique, called surface plasmon fingerprinting, had been developed for visualizing surface plasmon resonance with angular and spectral dependences.[21] In Ref. 21, the angular and spectral behavior of SPR was visualized using a tilting mirror system, and it was useful to evaluate the angular and spectral shift of SPR in order to understand the phenomena on the metal surface. It is desired to visualize the SPR fingerprints in the real time when one wants to evaluate reaction on the surface. Here we report that the spectroscopic attenuated total reflection (S-ATR) technique for real-time SPR visualization in which we employed the polychrometer and the two-dimensional CCD detector to monitor SPR fingerprinting in the real time.

2. THEORY

We will briefly explain an ordinary ATR image for SPR measurement before experimental results. Figure 1 shows the Kretschmann configuration in order to excite SPR.[3] Normally, thin metal film (~50 nm-thick Ag or Au) is located on the prism surface. In this configuration, the transmitted intensity I_T through the prism is calculated by the following equation

$$I_T=(1-R_1) R_2 (1-R_3) I_0 \quad (1)$$

, where I_0 is the incident intensity, and where R_1, R_2 and R_3 are the reflectivities at the plane in shown in Fig.1. As seen in equation (1), I_T is directly related with R_2. The relations between angles α, β and γ are expressed by the following equations

$$n_{air} \sin \alpha=n_{prism} \sin \beta \quad (2a)$$
$$\gamma=\beta+45 \ ^{\circ} \quad (2b)$$
$$n_{prism} \sin \gamma_c=1 \quad (2c)$$

, where n_{air} and n_{prism} are the refractive indices in the air and the prism depending on the wavelength λ of the incident light. $n_{air}=1$ for all the wavelength range. The total reflection occurred more than the critical angle γ_c. To simplify the situation, we consider that a 45 $^{\circ}$quartz (SiO$_2$) prism is located in the air and that an incident light at the wavelength of 633nm illuminates a 50 nm-thick Ag (or Au) film located on the prism. Since the critical angle γ_c=43.34 $^{\circ}$ for n_{prism}=1.457, the total reflection occurs γ>43.34 $^{\circ}$, β>-1.66 $^{\circ}$and α>-2.42 $^{\circ}$.

SPR can be excited only with P-polarized light, not with the S-polarized light. The SPR dispersion between air and metal surface is expressed by the following equations,

$$K_{SP}=(\varepsilon_{r,metal} \ \varepsilon_{r,air})/(\ \varepsilon_{r,metal} +\varepsilon_{r,air}) \quad (3a)$$
$$k_0= \omega/c=2\pi/\lambda_0 \quad (3b)$$
$$\varepsilon^*=\varepsilon_r+\varepsilon_i=n^* \quad (3c)$$

4

where K_x is the wave number of SPR along the metal surface, and ε^* and n^* are the complex dielectric constant and the complex refractive index of the material, respectively. k_0 and λ_0 are the wave number and the wavelength of incident light in vacuum, ω is the frequency and c is the light speed. Since the wavelength λ and the wave number k change in the material, ε^* is expressed as a function of ω like n^*.

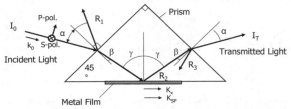

Fig.1. Kretschmann geometry of the ATR method in order to excite SPR. R_1 and R_3 are reflection at the prism surface, and R_2 is the reflectivity at the film.

Figure 2 shows the wavelength dependences of dielectric constants for silver (Ag), gold (Au) and quartz (SiO$_2$).[16] Normally, the real component of the dielectric constant is negative and the imaginary component is positive for metal, while the real component is positive and the imaginary component is almost zero for glass material. Since SPR has larger wave number compared with light, i.e. $K_x/k_0 > 1$, therefore $\varepsilon_r < -1$. As shown in Fig.2, ε_r of Ag is smaller than that of Au at the wavelength between 400 and 900 nm, which indicates that K_x/k_0 for Ag is smaller than that of Au in this wavelength range. For example, ε_r for Ag and Au at 633 nm are -17.46 and -9.62, which correspond to $K_{SP}/k_0 = 1.061$ and 1.117, respectively. The peak incident angle α for Ag must be smaller than that of Au. The reflectivity due to SPR for Kretschmann configuration is calculated including the thickness of the metal film as well as that of the coated layer on the metal film.[5]

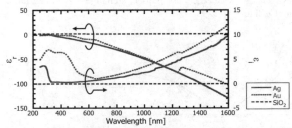

Fig.2. Dielectric constants used for the calculation. The real parts of the dielectric constants of metal films are normally negative.

In the ATR method using a prism to excite SPR, the wave number of light has to agree with the wave number K_{SP}. (See Fig.1) The wave number along the prism surface K_x is expressed by the following equation

$$K_x = (n_{prism} \sin \gamma) \, k_0 \quad (4)$$

In order to achieve above K_x/k_0, $n_{prism} \sin \gamma > 1$, i.e. $\gamma > \gamma_c$. Therefore SPR will appear in the total reflection region.

Figures 3 show calculated S-ATR images passed through the configuration shown in Fig. 1. SPR absorptions are clearly seen in the total reflection region in these figures. A 50 nm-thick Ag film in Fig.3 (a) shows the peak incident angle at the incident angle of 0.1 ° for the wavelength of 633 nm. In Figs.3 (a)-(c), the SPR absorption curves are shifted toward the right direction, i.e. the larger incident angle, as the thickness of the quartz coating increases. However, 15.5 nm-thick quartz (SiO$_2$) coated 50 nm-thick Ag film shows the same peak angle at 2.5 ° with the 50 nm-thick gold film for this wavelength. However, the S-ATR image for Au film shown in Fig.3 (d) can be easily distinguished from the S-ATR image for Ag film shown in Fig.3 (b). Therefore, the S-ATR image is important for visualizing SPR for an unknown film.

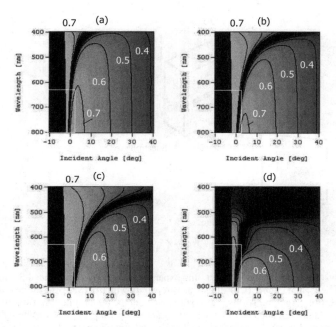

Fig.3. Calculated S-ATR Image for thin Au and Ag films. (a) 50 nm-thick Ag film, (b) 15.5 nm-thick and (c) 30 nm-thick SiO_2 coated Ag film as well as (d) 50 nm-thick Au film. The grayscale is colored for transmission I_T/I_0 from 0 (black) to 1 (white). Each contour indicates 0.1 differences in transmission. White corners in these figures show the incident angle of 2.5 ° and the wavelength of 633 nm.

3. EXPERIMENT

In order to visualize SPR, a two-dimensional CCD detector was used to detect SPR fingerprint in the coordination of the incident angle and the wavelength. Figure 4 shows geometry of the S-ATR technique. As shown in this figure, a white light source irradiatedwith wide incident angle toward a sample film located on a prism. The white light passed through the prism illuminates an entrance slit of a spectrometer in order to catch the angular distribution of the transmitted light that passed through the prism. The spectrometer is equipped with two-dimensional CCD detector. On the detector, each position of the image means a reflection property for a certain incident angle and a certain wavelength, indicating the angular-resolved and spectral-resolved ATR image, i.e. S-ATR image. The S-ATR image includes the dispersion property of SPR in the angle and the wavelength separately.

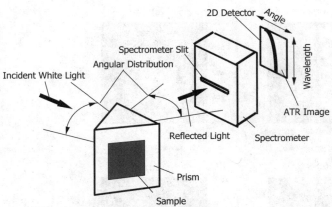

Fig.4. Schematic of the ATR imaging technique.

Figure 5 shows the experimental setup in this experiment. Vacuum evaporated 50nm-thick silver (Ag), gold (Au), and copper (Cu) films were prepared on quartz (SiO_2) substrates. A partially oxidized silver film (AgO_x on Ag) was also prepared from the 50 nm-thick Ag film by a plasma reactor. A 200 W tungsten lamp (I-150, CUDA Products Corp.) was used as a white light source. The light that passed through a polarizer (P) and through an achromatic lens (L1, f=80 mm) illuminates a pair of 45 ° quartz prisms used in order to straighten the incident beam toward the spectrometer. An achromatic lens (L2, f=80 mm) correct the dispersed light from the prism pair into the horizontal slit (S1, 100 μm-wide). A dove-prism and two achromatic lenses (L3 and L4, f=200 mm) were used to rotate the reflected light in order to send it into the vertical entrance slit (S2, 100 μm-wide) of the spectrometer. A spectrometer (Monospec 27, Jerral Ash Inc., f=270 mm) with a 150 lines/mm grating was used to diverse the transmitted light. A CCD camera (CS3330, Tokyo Electric Industries Co., Ltd.) is used as a two-dimensional detector. On the CCD image, the wavelength range from 476 to 782 nm was determined extrapolating the three He-Ne laser light at 543.5, 594.0, 632.8 nm, and the incident angle range from –3.1 ° to +6.1 ° was determined using a hard aperture between the lens (L1) and the 1st prism. The angular and spectral accuracy is 0.1 ° and 1.5nm, respectively.

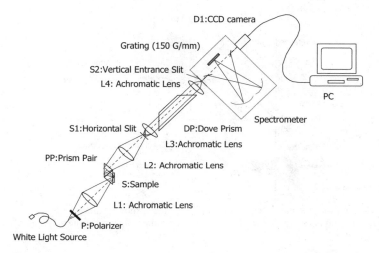

Fig.5. Schematic of the experimental setup is shown. A Dove prism (DP) is used to rotate the image of the horizontal slit (S1) in order to send it to the entrance slit (S2).

Figure 6 shows the details of the sample geometry consisting of two prisms and the substrate, which were optically connected with an index matching liquid. The transmitted intensity through this geometry system is expressed as

$$I_T = (1-R_1)\, R_{2,1st}\, R_{2,2nd}\, (1-R_3)\, I_0 \quad (5)$$

When the total reflection condition, equation (5) is same as equation (1).

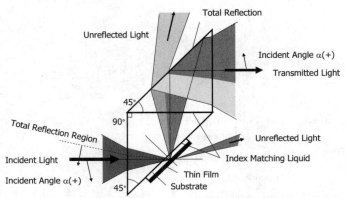

Fig.6. Top view of the pair of prisms and the substrate is shown. Index Matching Liquid is used between the two prisms and the sample substrate.

4. DISCUSSION

Figure 7 shows comparisons between a measured and a calculated S-ATR images for a 50 nm-thick Ag film with the P-polarized light. The SPR absorption is clearly observed. Dark SPR absorption curve shows good agreement.

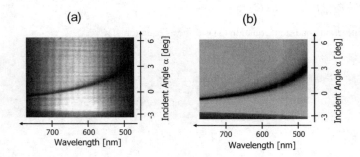

Fig.7. (a) Measured and (b) calculated S-ATR images for a 50nm-thick Ag film with the P-polarized light are shown.

Figures 8 show the S-ATR images for various films for the P-polarized and S-polarized light. SPR absorptions are observed only for the P-polarized light shown in the upper images. No absorption in the total reflection region is observed for the S-polarized light in the lower images. The intensity distribution along the wavelength would be derived from the geometrical condition or the intensity fluctuation of the incident light.

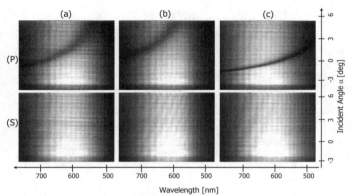

Fig.8. Measured ATR images for (upper, P) P-polarized and (lower, S) S-polarized from (a) Au, (b) Cu, and (c) Ag films. All films were 50nm thick.

Figure 9 show the normalized S-ATR images, which are calculated by the P-polarized images divided by the S-polarized images in order to normalize the intensity distribution. After this treatment, the normalized S-ATR images are clearly visualized. In Figure 9(b), the 50 nm-thick Au film has absorption peak near the incident angle of 2.5 ° and the wavelength of 633 nm. The 50 nm-thick Ag film shown in Fig.9 (c) shows a smaller peak angle at this wavelength, while the partially oxidized Ag film shown in Fig.9 (d) and the 50 nm-thick Cu film show a larger peak angle.Transmission is not small in these experimental results compared with the calculated S-ATR image shown in Fig.3. Better angular and spectral resolution will be required in order to evaluate film conditions quantitatively. However, in a specific situations such as deposition or oxidation process, the peak position will give useful information about the situationof metal surface.

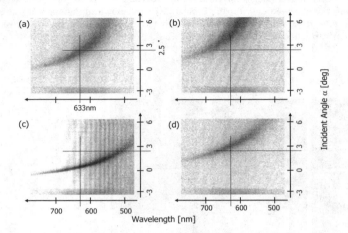

Fig.9. Normalized S-ATR images are shown for (a) Au, (b) Cu, (c) Ag as well as (d) partially oxidized Ag films.

5. CONCLUSIONS

In this paper, the spectroscopic attenuated total reflection (S-ATR) technique for real-time visualization of surface Plasmon resonance (SPR) is explained and demonstrated. Absorptions due to SPR on thin metal films are clearly visualized in the coordination of the incident angle and the wavelength. Metal films are distinguished even in partially oxidized condition. The real-time evaluation of SPR using this technique would be useful for monitoring biochemical reactions as well as for fabricating plasmonic devices.

ACKNOWLEDGMENTS

This work is partially supported by Bunkoh-keiki Co.,Ltd. (http://www.bunkoukeiki.co.jp)

REFERENCES

1. H. Raether, Surface Plasmons on Smooth and Rough Surfaces and on Gratings (Springer, Berlin, 1988).
2. A. Otto, Z. Phys. **216** (1968) 398.
3. E. Kretschmann, Z. Phys. **241** (1971) 313.
4. W. P. Chen and J. M. Chen, J. Opt. Soc. Am., **71** (1981) 189.

5. K. Kurosawa, R. M. Pierce, S. Ushioda, J. C. Hemminger, Phys. Rev. B **33** (1986) 789.
6. W. H. Weber and S. L. McCarthy, Appl. Phys. Lett. **25** (1974) 396.
7. W. H. Weber and S.L. McCarthy, Phys. Rev. B **12** (1975) 5643.
8. W. H. Weber, Phys. Rev. Lett. **39** (1977) 153.
9. Y. Aoki, K. Kato, K. Shinbo, F. Kaneko, T. Wakamatsu, IEICE Trans. Electron., **E81-C** (1998) 1098.
10. H. Liberdi and H. P. Grieneisen, Thin Solid Films **333** (1998) 82.
11. Y. Dai, T. Okamoto, I. Yamaguchi and M. Iwaki, Jpn. J. Appl. Phys. **32** (1993) L1269.
12. K. Saito, S. Honda, M. Watanabe and H. Yokoyama, Jpn. J. Appl. Phys. **33** (1994) 6218.
13. J. Raty, K. E. Peiponen, A. Jaaskelainen, M. O. A. Makinen, Appl. Spectrosc. **56** (2002) 935.
14. W. Hickel and W. Knoll, J. Appl. Phys. **67** (1990) 3572.
15. C. E. H. Berger, R. P. H. Kooyman, and J. Greve, Rev. Sci. Instrum. **65** (1994) 2829.
16. Handbook of optical constants of solids, edited by Edward D. Palik (Academic Press, Inc., Orlando, 1985) p.284, p.355, p.759.
17. X.-M. Zhu, P.-H. Lin, P. Ao, L.B. Sorensen, Sens. Actuators B **54** (1999) 3.
18. G. Steiner, V. Sablinskas, A. Hubner, Ch. Kuhne, R. Salzer, J. Mol. Structure **509** (1999) 265.
19. I. Stemmler, A. Brecht, G. Gauglitz, Sens. Actuators B **54** (1999) 98.
20. M. J. O'Brien, II, V. H. Prez-Luna, S.R.J. Brueck, G.P. Lopez, Biosens. Bioelectron. **16** (2001) 97.
21. M. Zangeneh, N. Doan, E. Sambriski, and R. H. Terrill, Appl. Spectroscopy **58** (2004) 10.

Mater. Res. Soc. Symp. Proc. Vol. 1182 © 2009 Materials Research Society   1182-EE09-21

Low-Energy Ion Beam Synthesis as a New Route Toward Plasmonic Nanostructures

Robert Carles, Cosmin Farcau, Julien Campos, Caroline Bonafos, Gérard Ben Assayag, and Antoine Zwick

CEMES – CNRS - Université de Toulouse, 29 rue Jeanne Marvig, 31055 Toulouse, France

ABSTRACT

Single δ-layers of dispersed silver (Ag) nanoparticles are obtained by low-energy ion beam implantation in a silica thin film. TEM microscopy reveals that the obtained Ag particles are spherical, crystalline, and the particles layer is located at only few nanometers below the free silica surface. We use reflectivity measurements to probe the optical/plasmonic response of the fabricated structures and exploit plasmon resonance and optical interference effects in the silica film to record the Raman scattering by quadrupolar vibrations of the spherical particles.

INTRODUCTION

Metallic nanostructures embedded in thin dielectric films are promising candidates for photonic/plasmonic devices. These systems can show interest for light guiding, confinement, and amplified optical phenomena thanks to the localization, enhancement and control of electromagnetic fields [1,2]. The most efficient metals for optical field enhancement in the visible or near-infrared range are silver (Ag), gold or copper, among which, the sharpest plasmon resonance is observed for silver [3]. Many effective ways have been explored to fabricate Ag nanocrystals (NCs), but the available techniques fail in satisfying all the requirements and the constraints in nanotechnologies such as the stability, sensitivity, uniformity and low cost for commercial applications. In that spirit, we propose the ultra-low-energy Ag^+ ion beam implantation in silica as a new method for Ag NCs synthesis. The advantage of this method is the direct embedding of the nanoparticles in a transparent dielectric matrix which does avoid their dissemination and their alteration, while preserving their plasmonic properties. Local field effects should be maintained on the flat surface of the substrate, provided that both the depth and size of the particles remain in the nanometer range.

We report for the first time on the ion-beam synthesis of a single plane (2-D array) of Ag nanocrystals within a SiO_2 layer on a Si substrate. We show how the optical properties of the prepared structures can be controlled by rationally engineering the geometrical parameters on both the micro- and nano-scale: i) by adjusting the thickness of the SiO_2 layer (100-200 nm) we can optimize the field enhancement at the surface due to interference effects, for the desired wavelength, ii) by adjusting the amount of implanted Ag^+ ions and their kinetic energy we control in the nanometer range (2-5 nm) the size of the Ag NCs and the depth at which they are formed below the SiO_2 free surface.

Transmission Electron Microscopy (TEM) was employed to analyze the morphology, location, and size distribution of the synthesized NCs. Reflectivity spectroscopy was used to probe the optical properties of the composite $Ag:SiO_2/Si$ samples. Plasmon resonance conditions and optical interference amplification (due to a specific silica layer thickness) allowed us to record the low frequency Brillouin-Raman signal of the single layer of Ag NCs. The observed low frequency spectra are well interpreted in terms of resonant Raman scattering by quadrupolar vibrations of spherical Ag NCs, and the results agree with TEM results.

EXPERIMENT

Synthesis and morphological characterization

A SiO_2 thin film (100 nm or 240 nm thick) thermally grown on a Si substrate was chosen as the host matrix for the Ag^+ implanted ions. According to simulation results, the experiments were performed with the following values for the kinetic energy: 0.65, 1.5, and 3 keV. In the case of 3 keV energy, Ag doses of 1.17, 2.34 and 4.68 x 10^{15} ions/cm² were used, which are expected to produce 5, 10 and 20 atomic percent of Ag at maximum of the implanted profile. The Ag NCs nucleate during the implantation process and no annealing procedure is necessary for the phase separation, as contrary to previous observations on Si NCs fabrication [4].

The location, size distribution and density of the Ag-NCs were deduced from TEM observations. Cross-sectional and plane-view TEM specimens were prepared by the standard procedure of grinding, dimpling and Ar^+ ion beam thinning. TEM was performed on a field emission TEM, FEI Tecnai™ F20 microscope operating at 200 kV, equipped with a corrector for spherical aberration.

Optical characterization

Reflectivity spectra were recorded by means of a Horiba-Jobin Yvon Uvisel ellipsometer, working with an incident angle of 45 degrees. Some measurements were performed by means of a Cary 5000 UV-NIR Varian spectrometer, working in quasi-normal incidence (8 degrees).

Raman characterizations were performed with a T64000 Horiba-Jobin Yvon triple spectrometer. To record the low-frequency and very weak signal originating from the Ag NCs, the entrance set-up of the spectrometer has been specifically modified. The laser beam was focused on the sample under Brewster incidence and the scattered light was collected in normal direction through a confocal microscope objective. The 413 nm radiation of a Krypton laser was used for excitation.

For all the optical measurements, we always kept a non-implanted area on each sample as a reference.

DISCUSSION

At first, TRIM simulations have been used to find adequate values for the implantation experimental parameters, e.g. the required kinetic energy of the Ag^+ ions. Figure 1 shows the expected profile of implanted Ag atomic fraction with respect to the depth (distance from the free SiO_2 surface), for 0.65, 1.5, and 3 keV. The same atomic fraction of Ag, 20%, can be obtained by doses of 2, 3.2, and 4.68 x 10^{15} ions/cm², in the three cases, respectively. An important observation is that the projected range R_p at which the maximum of the implantation profile is obtained can be controlled by the ion kinetic energy: It increases from 3 to 6 nm when increasing the energy from 0.65 to 3 keV. We expect therefore to be able to control on the nanometer scale the distance at which Ag NCs should be formed below the SiO_2 surface, by adjusting the implantation energy.

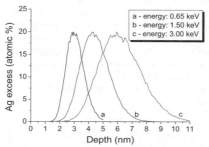

Figure 1. Implanted depth profiles for low-energy implantation of Ag^+ in SiO_2, obtained by TRIM simulations.

In Figure 2 representative TEM images are displayed for a series of samples prepared with the same energy 3 keV, but with different fraction of implanted Ag^+ ions (5, 10, and 20% at). Both cross-sectional (upper) and plane-view (lower) images recorded under Bright Field conditions are presented. Cross-sectional specimens reveal the presence of spherical Ag particles homogeneously distributed in a single plane. The particles are crystalline and made of pure silver, as determined by high resolution TEM imaging (not shown here). The optical spacing between the surface of the Ag NCs and the free surface remains remarkably constant and stands between 2.5 and 3 nm.

Figure 2. Cross- sectional (upper row) and plane-view (lower row) TEM images of samples implanted with 2, 3.2, and 4.7 x 10^{15} ions/cm^2 (from left to right) at 3keV kinetic energy.

The theoretically predicted position of the implanted profile maximum is compared with the experimentally determined position (relative to the free SiO_2 surface) of the Ag NCs in Figure 3A. A good agreement is obtained, although one observes that the experimental values are systematically larger than the predicted ones. This could be attributed to the matrix evolution during the implantation, in particular the expansion of the SiO_2 layer due to matter (Ag) addition, which is not taken into account in the simulation. In Figure 2B one observes that the mean diameter Φ of the particles increases from 1.5 to 2.9 nm by increasing the ion dose, while the size distributions show small dispersion. Plane-view observations allowed the evaluation of the nanoparticle density which varies from 6×10^{11} cm^{-2} to 10^{12} cm^{-2} when increasing the Ag excess from 5 to 20%. The mean interparticle distance varies accordingly in the range 13 to 10 nm, respectively (see Figure 3C).

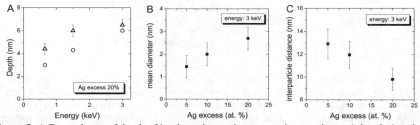

Figure 3. **A** Dependence of depth of implantation on ion energy in experiment (triangles) and simulation (circles). **B** Dependence of particle mean diameter on Ag dose. **C** Dependence of interparticle mean distance on Ag dose.

One interest was to validate non-destructive optical methods for the characterization of the new materials. Since the samples are on a Si substrate, and thus opaque, the optical response has been deduced from reflectivity measurements. Figure 4A shows the comparison between the response recorded from implanted (3 keV, 20%) and non-implanted sample areas. As we deal with a single plane of dispersed Ag-NCs, *i.e.* a volume lower than that of a 1 nm-thick layer, its presence is evidenced by a small shift and decrease of the reflectivity. All the reflectivity spectra display a pronounced dip around 500 nm which originates in a bilayer interference effect. The overall spectrum modification is a result of both a slight increase in the thickness of the layer, and a change in the refractive index, as we deduced by modeling based on Fresnel equations and a Maxwell-Garnett formula (results not shown here).

In very recent experiments we attempted to improve the optical interference effects by optimizing the sample structural parameters: the SiO_2 layer thickness has been modified to 240 nm to get $3\lambda/4$ equivalent thickness. Moreover the amount of implanted Ag ion dose has been increased to 9.4×10^{15} ions/cm^2. As a consequence, on one hand, a more pronounced dip develops in the reflectivity spectrum near 500 nm, and on the other hand, a shoulder around 410 nm clearly evidences the surface plasmon resonance (see Figure 4B).

It is well known that the surface plasmon resonance does occur around 410-420 nm in spherical and isolated silver nanoparticles embedded in glass [5]. Therefore the 413 nm radiation of a Krypton laser was chosen for Raman measurements, in order to take benefit of resonance effects on inelastic scattering cross section. Moreover, bilayer optical interference effects in the silica layer can further boost the intensity of the recorded signal [6]. The "dark-field" geometry

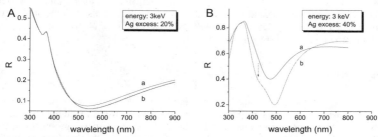

Figure 4. Reflectivity spectra of non-implanted (a) and implanted (b) SiO_2 / Si substrate. SiO_2 thickness is 100 nm in A and 240 nm in B.

of our modified setup ensures a very high rejection level of the spurious scattered light in the low-frequency range, allowing us to approach the Rayleigh line to 3-4 cm^{-1}. Using all these specific conditions Brillouin (Si substrate) and Raman (NCs) signals have been for the first time simultaneously recorded in a few seconds. An example is given in Figure 5, for the sample made with 3 keV implantation energy, and a silver excess of 20%. For comparison also the spectrum recorded in the same conditions from the non-implanted reference region is presented (b-black line). The sharp band centered at 7 cm^{-1} is due to Brillouin scattering from the Si substrate, while the broader band centered at 15 cm^{-1} is the Raman signature of Ag NC quadrupolar vibrations. The position of the latter strongly dependent on the sphere diameter, through the simplified relation: $\sigma(cm^{-1}) = 47/\Phi$ (nm), where the factor 47 depends on the type of vibration, the ratio of the longitudinal and transversal sound velocities, and the speed of light in vacuum [7]. The position of the measured Raman band yields a value of about 3 nm for Φ, which is in good agreement with the diameter of the NCs determined from TEM analyses.

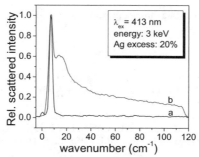

Figure 5. Low-frequency Brillouin-Raman scattering of non-implanted (a) and Ag implanted (b) SiO_2 layer on Si substrate.

It is very worthy of consideration the fact that by low-frequency Raman scattering we can determine the distribution of sizes of the nanoparticles, which is not straightforwardly possible with more conventional optical methods, like dark-field scattering or UV-visible extinction spectroscopy. This is because for Ag nanoparticles of spherical shape the plasmon resonance spectral position is almost independent on the particle size [5]. Note also that the observation of the Raman signature of such a low amount of Ag NCs has been possible by using an appropriate design of the structure and plasmon resonance effects.

CONCLUSIONS

Low energy ion implantation of Ag^+ within SiO_2 matrices is shown to be a relevant response for most of the challenges linked to the fabrication of efficient materials for plasmonic applications. TEM microscopy showed that the obtained Ag nanoparticles are spherical, crystalline, and dispersed in a planar array at few nanometers below the SiO_2 surface. We have evidenced the surface plasmon resonance of the NCs in reflectivity measurements, and exploited it to record the low-frequency Raman response of the NCs.

Several advantages offered by the proposed structure deserve to be briefly mentioned. Ag NCs are synthesized in one step at room temperature, embedded in a dielectric matrix, and therefore protected from contaminant adsorption. The SiO_2 surface is preserved optically flat and well adapted to work in reflection geometry, which is used for most applications both in the lab but also in field applications, i.e. in portable optical fiber-based spectrometers. The fabrication process is compatible with industrial wafer-scale Si technology and can be thus combined with various techniques. It can be implemented at different steps of a complex lithographic procedure

Very importantly, the distance between the Ag NCs and the free SiO_2 surface is kept constant over large areas, well controlled in the 1-5 nm range, *i.e.* in the desired range to exploit near-field optical coupling between nano-objects (plasmonic antenna, molecules or quantum dots).

ACKNOWLEDGMENTS

C.F. acknowledges funding through project BQR – PRES Univ. Toulouse 2008/2009.

REFERENCES

1. R. Sainidou, F. J. Garcia de Abajo, *Opt. Express* **16**, 4499 (2008).
2. A. Biswas, H. Eilers, F. Hidden, O. C. Aktas, C. V. S. Kiran, *Appl. Phys. Lett.* **88**, 013103 (2006).
3. W. A. Murray, W. L. Barnes, *Adv. Mater.* **19**, 3771 (2007).
4. C. Bonafos, M. Carrada, N. Cherkashin, H. Coffin, D. Chassaing, G. Ben Assayag, A. Claverie, T. Müller, K. H. Heinig, M. Perego, M. Fanciulli, P. Dimitrakis, P. Normand, *J. Appl. Phys.* **95**, 5696 (2004).
5. J. Lerme, B. Palpant, B. Prevel, M. Pellarin, M. Treilleux, J. L. Vialle, A. Perez, M. Broyer, *Phys. Rev. Lett.* **80**, 5105 (1998).
6. W. S. Bacsa, J. S. Lannin, *Appl. Phys. Lett.* **61**, 19 (1992).
7. H. Portales, L. Saviot, E. Duval, M. Fujii, S. Hayashi, N. Del Fatti, F. Vallee, *J. Chem. Phys.* **115**, 3444 (2001).

Mater. Res. Soc. Symp. Proc. Vol. 1182 © 2009 Materials Research Society   1182-EE09-38

Ferroelectric Thin Film Microcavities and Their Optical Resonant Properties

Pao Tai Lin[1], William A. Russin[2], Alexandra Imre[3], Leonidas E. Ocola[3] and Bruce W. Wessels[1,4,*]

[1] Department of Materials Science and Engineering, Northwestern University, Evanston, Illinois 62028
[2] Biological Imaging Facility, Northwestern University, Evanston, Illinois 62028
[3] Center for Nanoscale Materials, Argonne National Laboratory, Argonne, Illinois 60439, USA
[4] Electrical Engineering and Computer Science, Northwestern University, Evanston, Illinois 62028

ABSTRACT

Two dimensional photonic crystal (PhC) microcavity structures were fabricated using epitaxial ferroelectric thin film as the optical media and their resonant optical properties were measured. The PhC structures are utilized in order to achieve strong light localization to enhance the interaction between the incident light and the nonlinear optical barium titanate (BTO). Fluorescence measurements were used to assess the resonant properties. Two types of resonant structures were investigated consisting of either dopant or vacancy PhC arrays. The nano patterning on BTO thin films was achieved using dual beam focused ion beam (FIB). For the dopant type PhC microcavity structure, a larger air hole is generated in every 5 x 5 unit cells forming a super cell. The spatial profiles of PhC microcavity structures are characterized by laser scanning confocal microscopy. Structures with a feature size approaching the optical diffraction limit are clearly resolved. Fluorescence measurements on PhCs coated with a fluorescent dye were carried out to determine the relationship between the degree of light localization and the photonic band structure. Enhanced fluorescence at wavelengths 550-600 nm is observed in the dye covered PhCs with lattice period a =200 nm and a =400 nm. The large fluorescence enhancement results from the presence of PhC stop bands that increase the emission extraction efficiency due to the strong light confinement. Since the only allowed propagation direction for the scattered fluorescence light is out-of-plane, this enhances the vertically fluorescent extraction efficiency. These BTO optical microcavity structures can potentially serve as active nano-photonic components in bio-sensors and integrated photonic circuits.

INTRODUCTION

Photonic crystals based on non-linear materials have been considered as a promising candidate for integrated optics circuits due to their enhanced nonlinearity and compact size[1]. PhCs containing microcavities are especially interesting for nano-photonic devices since their high quality Q factor enhances interaction between the materials and photons. Narrow bands are generated inside the photonic band gap by introducing optical cavities[2]. These microcavities serve as defects that trap photons. Significant enhancement of non-linearities and light emission has been demonstrated by using of artificial photonic lattices[3]. Recently, manipulating light emission by forming PhC from photoluminescent materials has attracted much attention for its application of fluorescence quantum dots and color controlled protein visualization[4]. For many of these applications ferroelectric oxides such as lithium niobate and barium titanate (BTO) are especially promising since they have large non-linear optical coefficients[5]. Furthermore, BTO is transparent in the visible to mid-IR spectral region that makes it a candidate for ultra-broad band integrated photonics[6]. Recently we have investigated PhC fabricated using BTO and have demonstrated a hybrid BTO thin film waveguide structure where the PhC is patterned in the silicon nitride cladding layers[2]. A Bragg grating 300 micron long with 27 nm wide stop band at $\lambda=1.55$ μm was obtained. To further decrease the interaction length, a much higher refractive index contrast between the high and low dielectric regions is necessary that requires direct patterning of the BTO layer. However, difficulties in patterning barium titanate potentially limit its application in nano-photonic components due to its chemical inertness and mechanical hardness.

To overcome the challenges of nano patterning of BTO, we use focused ion beam milling. The direct patterning of PhC with microcavities in epitaxial ferroelectric oxide thin films has achieved where a feature size as small as 100 nm and a high aspect ratio of 2.5 have been obtained. In this report the optical properties of the photonic microcavities were characterized using confocal microscopy, which is especially useful in the characterization of structures with a feature size approaching the optical diffraction limit. The confocal microscopy images show that the optical response is determined by the symmetry of the embedded microcavity arrays. To evaluate the photonic band structures in the visible region, the fluorescence spectra of the dye covered BTO PhC were measured. Strongly enhanced fluorescence was observed that indicates improved emission extraction efficiency was obtained. The enhancement is attributed to the forbidden in-plane wave propagation due to the presence of a photonic band gap. Hence the work suggests that the sensitivity of fluorescent bio-sensors and the detectors can be significantly improved by optimizing the PhC structures.

EXPERIMENTAL & DISCUSSION

Fabrication of BTO PhC Cavities

PhCs were fabricated from epitaxial BTO thin films prepared by metalorganic chemical vapor deposition (MOCVD) on an MgO substrate as described elsewhere[3]. The BTO layer was 500 nm thick. The PhC was patterned by focused ion beam milling (FEI Nova 600 NanoLab Dual Beam). A 20 nm thick Au layer is deposited on the sample prior to the ion milling in order to avoid the insulator charging. The ion beam current was 93 pA and the dwell time of each milled pixel is 500 ms. Fig. 1 shows a schematic of the PhC cavity structure used in this study. The PhC consists of a square array of air holes with a lattice constant **a** ranging from 200 nm to 1 μm. Each array contains 24 X 24 unit cells. The active length was 5 to 24 micron. Two types of PhC cavity structures were fabricated by incorporating defects. Figure 1(a) shows the schematic of the dopant type PhC cavity structure, where a defect consisting of a wider air hole is generated in every 5 x 5 unit cells forming a super cell. They serve as the photonic dopant and the photonic super cell, respectively. In the case of the vacancy type PhC cavity a missing air hole acts as a vacancy. There is an air hole missing in every five PhC unit cells as shown in figure 1(b). Figure 1 (c) is the side view of the desired BTO PhC cavity structure, where **a** is the lattice constant, h_{BTO} is the film thickness, and $2r$ is the diameter of the milled hole. The terms r_d and r_v are the hole radii of dopant (enlarged hole) and vacancy (missing hole), respectively. The milled hole depth is the same as the BTO film thickness $h_{BTO}=500$ nm. Figure 1 (d) is the top view of desired PhC unit cell. The gray area is BTO region, which has high dielectric constant and refractive index n_{BTO}=2.3. The white area is the low dielectric air region with n_{air}=1. A high refractive index contrast Δn=1.3 is obtained.

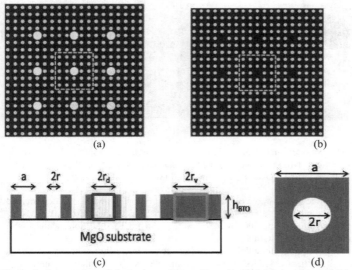

(a)   (b)

(c)   (d)

Figure 1. Schematics of photonic cavities with embedded (a) dopants and (b) vacancies. The white circular array consists of air holes and gray background is BTO. The yellow square indicates a photonic supercell. (c) Side view of the designed PhC cavities. The red and the green squares indicate a photonic dopant and vacancy site. The terms a, $2r$, $2r_d$, $2r_v$ represent the photonic lattice constant, air hole diameter, dopant diameter, and vacancy diameter, respectively. (d) Top view of a photonic lattice.

Topography of BTO PhC Cavities

The band structure of fabricated PhC is determined by the lattice symmetry and the defect structure. The structure of BTO PhC was measured by scanning electron microscopy (SEM). Figure 2 (a) shows the cross sectional image of PhC lattices and an enlarged image is shown in Figure 2 (b). The dark gray columns are the ion milled air holes and the light gray background is the BTO dielectric. The milled holes are uniformly distributed inside the patterned area. From the image in Figure 2 (b), the measured photonic lattice constant a is 400 nm, where the milling depth h_{BTO} and the hole diameter $2r$ are 500 and 400 nm, respectively. The hole is wider at the top opening due to the ion beam having a Gaussian beam profile. While there is some variation of the hole radius in the direction perpendicular to film surface, the PhC periodicity in the thin film plane is still well-maintained. From the SEM image the change of periodicity is less than 20 nm over the entire patterned area. No distortion is found in the milled hole, which indicates that stigmatism in the focused ion beam optics is not important.

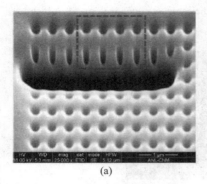
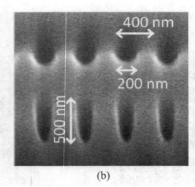

(a)   (b)

Figure 2. (a) SEM cross-sectional images of the fabricated BTO photonic crystals. An enlarged image of the area inside the square array outlined in red is shown in (b). The milling depth, air hole diameter, and photonic lattice constant are 500nm, 200 nm, and 400 nm, respectively.

Figure 3 (a) shows the top view (32 degree tilted) of the dopant type PhC, where the dopants are uniformly distributed inside photonic lattice arrays. From figure 3 (a) the periodicity Λ (=a) and the air hole diameter $2r$ are 400 nm and 200 nm, respectively. An enlarged air hole, which serves as the photonic dopant, is embedded with a period of 5Λ. The dopant hole diameter $2r_d$ is 350 nm. Structures consisting of optical super cells containing vacancies are shown in figure 3 (b), which displays the 4 X 4 cavities embedded in 24 X 24 unit cells. At the vacancy site there is a missing air hole. From figure 3 (b) the measured vacancy diameter $2r_v$ is determined to be the same as the lattice constant a=400 nm.

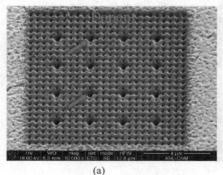
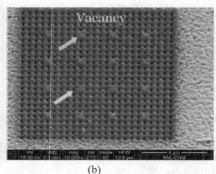

(a)   (b)

Figure 3. SEM top image of the fabricated BTO PhC cavities. A 4 x 4 array of photonic dopants are embedded in a 24 x 24 PhC. (a) The purple arrows indicate the sites of the photonic dopants which shows the dopant cavity has a opening of 350 nm. (b) The blue arrays indicate the sites of the photonic vacancies which tell the dopant vacancy diameter is 400 nm.

Enhanced photoluminescence of dye covered BTO PhCs

To study the effect of PhC structure on scattered light intensity, fluorescence images from a dye coated PhCs were investigated. Photonic band structures are revealed by analyzing the variation of light emission efficiency for PhCs with different lattice periods Λ. The fluorescence extraction will significantly increase when the in-plane light propagation is prohibited due to the presence of photonic bands gaps. Meanwhile the band gap center and width are mainly determined by lattice symmetry and Λ. The fluorescence from the photonic microcavities imaged by a laser scanning confocal microscope are shown in figure 4. The emission range of the dye is from 515 nm to 625 nm with the emission peak at 550 nm and the FWHM of the emission spectrum is 35 nm. The excitation light was filtered out. The captured fluorescence images of the PhC cavities are shown in figure 4. The yellow and brown colors indicate strong and weak fluorescence, respectively. In this figure there are seven separate PhCs with different lattice constants. The lattice period Λ for the seven PhCs are 200 nm, 300 nm, 350 nm, 400 nm, 450 nm, 500 nm, and 550 nm, respectively. Significant fluorescence enhancement is observed for the PhC with Λ=200 nm, which has a golden yellow color. Minor fluorescence enhancement is also found for Λ=400 nm. On the other hand, the PhC with Λ=300 nm is brown which indicates a decrease of fluorescence.

The differences in fluorescence response of the PhC structures can be explained by the presence of a stop band. If the dye emission spectrum coincides with the stop band, the fluorescence is enhanced due to the in-plane light propagation being prohibited. The only allowed propagation direction for the scattered fluorescence light is out-of-plane (vertical) emission. In other words, the light emitted is concentrated and projected only toward a small portion of an emission sphere. Therefore the increased fluorescence indicates the light confinement caused by PhCs array intensifies the fluorescence extraction efficiency. Desired photonic band gaps can be achieved by choice of the photonic lattice constants **a**. It should be noted that the fluorescence excitation level is in the same layer with the PhCs pattern. The focal plane used to collect the light emission is maintained at the same height during the entire experiments.

The presence of an optical supercell has a considerable effect on the fluorescence as shown in figure 4, where the dopant sites are indicated by white arrows. It can either increase or decrease the fluorescence efficiency. The effect of PhC cavity on intensified fluorescence mainly depends on two factors: the optical lattice period and the light interference from the supercell. For optical dopants in PhC with Λ=400 nm, a decreased fluorescence intensity is readily observed at the dopant sites compared to the neighboring PhC lattices. Decreased fluorescence emission from cavities is also found in figure 4 for PhC with Λ=200 nm. The decrease in fluorescent is due to the formation of a narrow transmission band inside the original photonic stop band. The new band is generated by the cavity arrays that break the symmetry of PhC. Thus, the original confinement of fluorescence by photonic stop bands no longer occurs and as a consequence light can emit from a wider solid angle. On the other hand, optical dopants in PhC with Λ=300 nm have a weakly enhanced fluorescence in contrast to their neighboring photonic lattices. This may be due to a constructive light scattering, which has been discussed in last section. For other PhC cavities with larger lattice constant Λ, there are only negligible effects on fluorescence because the photonic stop bands are far away from the dye's emission region.

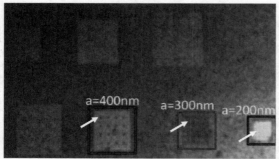

Figure 4. Fluorescence images of BTO PhCs cavity structures. The excitation light is He/Ne 543 nm laser. The lattice period Λ of the PhCs are 350 nm, 450 nm, 550 nm, 650 nm, 500 nm, 400nm, 300nm, 200 nm from left to right in the top to bottom rows, respectively. The dopant sites are indicated by white arrows.

CONCLUSIONS

In conclusion, PhC supercells using epitaxial BTO thin film are demonstrated. Two types of PhC macro-cavities including optical dopant or vacancy defects were fabricated in the high optically nonlinear BTO. Using FIB a high resolution pattern of 100 nm and large aspect ratio of 2.5 are achieved. To study the optical resonance of the PhC microcavities confocal microscopy is used. Photonic band structures were also characterized by measuring the fluorescence of the dye covered PhCs. Strongly enhanced fluorescence is obtained for PhCs at Λ=200 nm. This significant enhancement is contributed by the selective angular fluorescence emission since the in-plane wave propagation is prevented due to the presence of a photonic band gap. There is a clear overlap between the optical stop bands and the fluorescence spectra which intensifies the vertical emission. This study paves the ways of using BTO optical cavities as active nano-photonic components and bio-sensors.

ACKNOWLEDGMENTS

This work was supported by the National Science Foundation through ECCS Grant No. 0801684, the NSF MRSEC program through the Northwestern Materials Research Center (DMR-0076097) and Argonne national laboratory Center for nanoscale materials No. 282.

REFERENCES
1. J. D. Joannopoulos, S. G. Johnson, J. N. Winn, R. D. Meade: Photonic Crystals: Molding the Flow of Light, 2nd Edition, Princeton, New Jersey: Princeton University Press; 2008
2. Z. Liu, P. –T. Lin, B. W. Wessels, F. Yi and S. –T. Ho, Appl. Phys. Lett. **90**, 201104 (2007)
3. P. T. Lin, Z. Liu and B. W. Wessels, J. Opt. A: Pure Appl. Opt., accepted (2009)
4. A. Badolato, K. Hennessy, M. Atatüre, J. Dreiser, E. Hu, P. M. Petroff and A. Imamolu, Science **308** 1158 (2005)
5. P. T. Lin, B. W. Wessels, J. I. Jang and J. B. Ketterson, Appl. Phys. Lett. **92**, 221103 (2008)
6. P. Tang, D. Towner, T. Hamano, A. Meier and B. Wessels, Optics Express, **12**, 5962 (2004)

Metamaterials and Superlenses

Mater. Res. Soc. Symp. Proc. Vol. 1182 © 2009 Materials Research Society   1182-EE11-02

Improved Analytical Models for Single- and Multi-Layer Silver Superlenses

Ciaran Moore[1], Richard J. Blaikie[1] and Matthew D. Arnold[2]

[1] MacDiarmid Institute for Advanced Materials and Nanotechnology, Department of Electrical and Computer Engineering, University of Canterbury, Private Bag 4800, Christchurch, N.Z., email for correspondence: richard.blaikie@canterbury.ac.nz

[2] Institute for Nanoscale Technology, Department of Physics and Advanced Materials, University of Technology Sydney, PO Box 123 Broadway, NSW 2007, Australia

ABSTRACT

Spatial-frequency transfer functions are regularly used to model the imaging performance of near-field 'superlens' systems. However, these do not account for interactions between the object that is being imaged and the superlens itself. As the imaging in these systems is in the near field, such interactions are important to consider if accurate performance estimates are to be obtained. We present here a simple analytical modification that can be made to the transfer function to account for near-field interactions for objects consisting of small apertures in otherwise-continuous metal screens. The modified transfer functions are evaluated by comparison with full-field finite-element simulations for representative single-layer and multi-layer silver superlenses, and good agreement is found.

INTRODUCTION

The resolution of an optical imaging system is fundamentally limited by diffraction, and the consequences are well understood for conventional far-field systems. By working in the near-field region, evanescent waves, which contain high spatial-frequency information, can also contribute to the image and conventional diffraction limits can be overcome. Various scanned-probe optical microscopy systems [1], or mask-based near-field lithography systems [2] have been developed and have been used to experimentally confirm such sub-diffraction-limited near-field resolution.

However, the exponential decay of evanescent fields usually means that the best resolution for a near-field image is obtained right at the object plane, and the concept of image projection is not easily translated from far-field to near-field imaging systems. This changed in 2000 with the publication of Pendry's seminal 'perfect lens' paper [3], where the concept of perfect or near-perfect imaging using negative-index materials was elegantly described. For near-field imaging systems this introduced the concept of using a planar plasmonic layer for super-resolved image projection, and subsequent experimental work using silver planar lenses has shown this to be a realizable effect [4,5].

These planar lenses rely on resonant plasmonic interactions to mediate the image formation and, as such, the transfer of spatial-frequency information from the object plane to the image plane is rich and complex. One common way to analyse this has been to use the spatial-frequency transfer function for the lens on its own, which is a straightforward and exact analytical calculation that has been described in many publications [6,7,8,9]. There are typically resonant peaks and notches in this spatial-frequency transfer function, and these lead to many object-dependent imaging artefacts that we have described previously [10]. In addition, the object and the planar lens are in the near-field zones of each other, so strong object-lens interactions may be expected. Such interactions cannot be easily and exactly incorporated into the simple transfer-matrix calculations, so they have not been studied in any great detail to date.

Here we show how object-supelens interactions can be accounted for approximately in a modified transfer function, and we compare analytical and finite-element simulation results to quantify the improved accuracy of our new method. Both single-layer and multi-layer planar lens systems are studied and we highlight the effects that object-lens interactions can have on the spatial frequency transfer function in each case.

THEORY AND MODELLING TECHNIQUES

To attain sub-diffraction limited imaging, superlenses rely on negative refraction or plasmonic resonances in one or more of their layers to enhance an object's evanescent near field [8]. An explanation of the development of this idea and the physical cause behind it is discussed in this section. Consideration is also given to the way that this phenomenon is handled by different modeling techniques.

Originally, negative permittivity, ε, and negative permeability, μ, were both thought to be prerequisites to achieve a negative index of refraction, $n = (\varepsilon\mu)^{-1/2}$ [11]. This limited the development of metamaterial science, as materials with $\mu < 0$ are not found in nature. Fortunately, further analysis of sub-wavelength systems revealed that the electrical and magnetic fields became decoupled when all features were much smaller than a wavelength, i.e. $d << \lambda$ [3]. This meant that negative refraction could be achieved with only negative ε, allowing metals, which have negative ε in the visible spectrum, to be used as superlensing media.

Later experimental work confirmed surface plasmon resonances (SPRs) at a metal/dielectric interface, such as silver/silicon dioxide, can be excited by transverse magnetic (TM)-polarized light of a suitable wavelength (~365 nm for silver) and used for superlens imaging [4,5]. SPRs occur at both metal interfaces, and are strongly coupled to each other provided the loss through the slab is minimal. SPRs on the second interface enhance the local electric fields there, leading to sub-diffraction limited imaging at some distance beyond the source.

In order to study this imaging process further and to improve the modeling of superlenses, two techniques were chosen and are described here. Firstly, the transfer-matrix (T-matrix) technique is examined. The T-matrix technique calculates transmission and reflection parameters for one-dimensional, layered systems, allowing ideal structures to be characterised in the spatial frequency domain. It does not accommodate discontinuous geometries but can estimate their performance in some cases.

Secondly, finite element modeling (FEM) was chosen to complement the T-matrix approach by working primarily in the spatial domain, generating two- or three-dimensional plots of the electromagnetic fields present in a superlensing system. It can cater for boundary conditions and non-ideal input sources and its results can be converted to the frequency domain, to allow direct comparison with T-matrix results. The disadvantage of this approach is increased complexity and computation time. The theory and practice behind both of these techniques are explained in the sections that follow.

Transfer-matrix (T-matrix) Theory

T-matrices describe the transmission and reflection characteristics of an interface between two media. Multiplying two or more T-matrices together allows the behaviour of multi-interface geometries to be calculated, provided appropriate phase shifts are applied to the results. We used this T-matrix approach to determine the spatial frequency transfer functions of several different superlenses, using the formalism shown in Figure 1 [9]. Here E is the electric field, H the magnetic, and k is the corresponding wavevector. Further details of this analysis can be found in the literature [9,10].

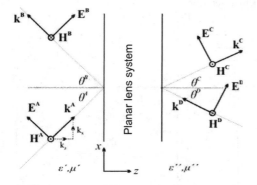

Figure 1. T-matrix simulation domain.

In this analysis the field components on the left and right sides of a set of planar interfaces, as depicted in Figure 1, are used to determine spatial-frequency transfer functions from 2×2 T-matrices, where the T-matrix elements are determined from boundary matching and phase-shift considerations [10]. For any particular wavevector parallel to the interfaces, k_x, the transmission and reflection coefficients for the system are determined from the matrix coefficients by

$$t = \frac{E_x^C}{E_x^A},$$ (1)

$$r = \frac{E_x^B}{E_x^A}.$$ (2)

Calculating transmitted and reflected intensity, $|t|^2$ and $|r|^2$, over a range of wavenumbers allows transfer and reflection functions to be constructed, giving two spatial-frequency-domain metrics for lens performance. By way of example, two superlenses are characterized here. The first superlens [3] is made of a 40 nm Ag layer sandwiched between two 20 nm SiO$_2$ layers, the properties of which are defined in Table I. The second superlens is more sophisticated, containing three layers of Ag, 13.3 nm thick, separated by two layers of SiO$_2$, also with thickness of 13.3 nm. The structure is capped at each end by a layer of SiO$_2$, half as thick as before at 6.7 nm [10]. Note that both lenses have the same total thickness of 80 nm, but the multilayer structure is predicted to have superior performance at high spatial frequencies [8]. The transmission and reflection functions for the single- and triple-layer lenses are shown in Figure 2. Applying the transmission functions to a test object (two 20-nm wide slits separated by 100 nm) [8] gives the spatial image profiles shown in Figure 3, highlighting the improvements that are expected for the multi-layer lens.

Table I. Material properties used for superlens modeling.

Material	ε_r	μ_r	Ref.
Ag	-2.7 + 0.23i	1	[12]
W	1.497 + 7.690i	1	[13]
SiO$_2$	2.368	1	[13]

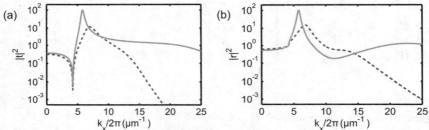

Figure 2. Transmission functions (a) and reflection functions (b) for single (dashed) and triple-layer (solid) Ag superlenses.

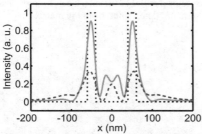

Figure 3. Spatial output for single (dashed) and triple (solid) Ag layer superlenses exposed to a double slit mask (dotted) [8].

Modified T-matrix Theory Including Mask Reflections

Although the T-matrix technique accurately models mode behaviour within a planar, continuous superlens, it cannot describe behaviour in the presence of discontinuous layers in front of or beyond the superlens. Specifically, T-matrices cannot account for mask-lens interactions, which arise between a superlens and the structured object that it images. These interaction effects may contribute significantly to superlens performance; hence an addition is made to the T-matrix technique for the first time here that allows this to be accounted for.

The addition begins by approximating the reflection function from a shadow object with the reflection function of a solid layer of mask metal, r_M. This metal layer, taken to be tungsten in the following examples, is placed some distance, d, from the lens, giving the domain shown in Figure 4. Next, the intensity profile seen at the input of the lens, which was originally ip, is affected by the resulting infinite loop of reflections between the mask and the lens, and becomes

$$ip_{new} = ip + ip \cdot r_L r_M e^{i2k_z d} + ip \cdot \left(r_L r_M e^{i2k_z d}\right)^2 + ip \cdot \left(r_L r_M e^{i2k_z d}\right)^3 + \dots \qquad (3)$$

where r_L is the lens reflection function. This changed input causes a change in the output of the lens, which becomes

$$op_{new} = t \cdot \left(ip + ip \cdot r_L r_M e^{i2k_z d} + ip \cdot \left(r_L r_M e^{i2k_z d}\right)^2 + ip \cdot \left(r_L r_M e^{i2k_z d}\right)^3 + \dots\right) \qquad (4)$$

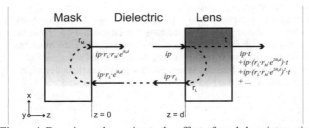

Figure 4. Domain used to estimate the effect of mask-lens interactions.

The modified transfer function of the lens can now be found by making three simplifications: firstly, ip is set to 1, removing it from the equation and making the output intensity profile normalised to a unit-intensity input (i.e. a transfer function). Secondly, the mask is put in contact with the lens (including spacer layers), forcing $d = 0$ and removing the exponential terms from the equation. Thirdly, the infinite sum in Eqn. 4 is collapsed, by noting that an infinite sum of terms, $\sum a^n$, converges to $1 / (1 - a)$. This allows the modified transfer function, which takes into account an estimate of the mask-lens interactions, to be given as

$$t_{new} = \frac{t}{1 - r_L r_M} \qquad (5)$$

Modified transfer functions for the lenses described in the previous section are shown in Figure 5 (b), with the unmodified transfer function given for comparison in Figure 5 (a). It can be seen that the mask-lens interaction changes the transfer function significantly, with the resonant peaks at 6-8 μm^{-1} being damped and moved to higher spatial frequencies.

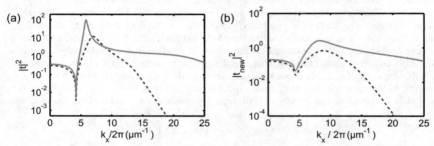

Figure 5. Original (a) and modified (b) transfer functions for single- (dashed) and triple-layer (solid) Ag superlenses taking into account a tungsten object layer.

Although t_{new} goes some of the way towards describing the interaction between the lens and the mask, there are two caveats for its results to be valid. The first stems from the assumption that a solid sheet of metal can approximate a mask. This assumption is relatively good when the mask itself is mostly metal, with only small gaps through which light can be transmitted. It is harder to justify this assumption when the exposed area is much greater than the area covered by the mask; however, in such a situation the mask-lens interactions will be much weaker and there will be less of a need to modify the transfer function. This leaves the case where the mask has a duty cycle close to 50%; here the need for a modification to the original transfer function is significant, however, the assumption used to find t_{new} is invalid. The modifications described above would not work in this case and a new method would need to be investigated.

The second restriction on t_{new} involves the validity of the infinite sum of reflection terms. Strictly, $|a| < 1$ is required for $\sum a^n = 1 / (1-a)$ to be valid. This appears to be a problem, as $|r_L r_M|$ is greater than 1 for a range of wavenumbers around the main resonance in the transfer function (Figure 6). Fortunately, full-field simulations, discussed in the next section, show that the physical meaning of breaking this requirement is less severe than the mathematical rules would suggest, hence the approximation will continue to be used until a better one can be found.

Finite-element Modeling

The COMSOL multiphysics engine [14] was used to construct finite-element models (FEMs) that allowed the spatial response of several thin-film planar lenses (superlenses) to be characterized. Each model described a two-dimensional plane, running through the component layers of a superlens. A separate sub-domain in the simulation plane, with explicitly defined geometry and electromagnetic properties, represented each superlens layer. An example of such a simulation domain, containing a superlens and a shadow mask object, is shown in Figure 7. In this example, the simulation domain measures 1 μm by 376 nm, with 10 nm wide perfectly

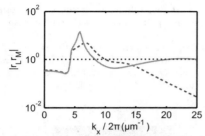

Figure 6. $|r_L r_M|$ for single- (dashed) and triple-layer (solid) Ag superlenses.

matched layers (PMLs) at the vertical boundaries. The left-most (source) boundary represents an electric field, with $\boldsymbol{E} = E_x = 1 \cdot e^{ik_0 z}$, where the wavelength of operation is 365 nm. The opposite boundary has a PML to minimise boundary-reflection effects. Horizontal boundaries are set to symmetric periodic conditions, which make the simulation domain infinitely periodic. The mask is made up of two pieces of tungsten, 40 nm thick and separated by a 50 nm gap. The active layer in the superlens is silver, also 40 nm thick and separated from the mask by a distance of 20 nm. All other subdomains are comprised of silicon dioxide. The electromagnetic properties of each of these materials are shown in Table I.

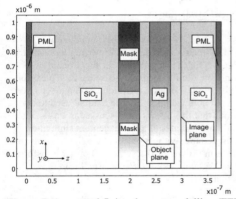

Figure 7. Annotated finite-element modelling (FEM) simulation domain.

Once specified, the simulation domain was divided into finite elements or mesh points, each of which was able to interact with its neighbours or one or more simulation boundaries. The number of elements, or mesh density, could be varied arbitrarily, with the maximum number of mesh points limited by the amount of computer memory available: roughly 1 GB of RAM was needed for every 100,000 mesh points. On a 32-bit operating system, this allowed for average mesh densities between 0.27 and 0.53 points/nm². Figure 8 (a) shows a set of mesh points around the mask feature from Figure 7. The mesh contains 162208 elements, giving an average mesh density of 0.43 points/nm². Simulation variables, such as the electric field, were calculated at each of the mesh points, with their values given by iterative solutions to the differential forms of Maxwell's equations. An example of such a solution is given in Figure 8 (b) for the mesh shown in Figure 8 (a).

The smoothness of the calculated fields depended, in large part, on the density of the underlying mesh. An insufficiently meshed domain would yield poorly-resolved field profiles that contained erroneous features. To avoid this, we ran a number of convergence tests to ensure that the mesh density of Figure 8 (a) was sufficient to provide accurate simulations.

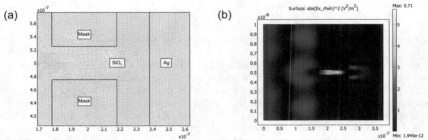

Figure 8. FEM model for the geometry shown in Figure 7. (a) Mesh detail; (b) Simulated electric field.

Transfer-matrix Estimation from Finite-element Modelling

Although FEM gives comprehensive field solutions, its results are not immediately comparable to the transfer functions found via the T-matrix technique. Converting the T-matrix results to the spatial domain, as was done in Figure 3 for the simple transfer function, allows such comparison to be made; however, this is only on a limited, case-by-case basis. To exploit both the data quality of FEM and the data quantity of the T-matrix technique, comparisons need to be made between transfer functions derived from both techniques. The method used to derive FEM-based transfer functions and the comparisons that this allows are discussed in this section.

Firstly, line scans are taken at the object and image planes of a finite element simulation. The model shown in Figure 7 is taken as an example here, with its object plane field, *op*, shown in Figure 9 (a) and its image plane field, *ip*, shown in Figure 9 (b). Next, the Fourier transforms of both of these fields are calculated to give input, *IP*, and output, *OP*, spectra. From these spectra, the transfer function for the lens is calculated as

$$t_{FEM} = \frac{OP}{IP} \qquad (6)$$

Figure 10 compares the FEM-generated transfer functions t_{FEM} (a) with the reflection-modified transfer functions t_{new} (b), and good agreement is observed.

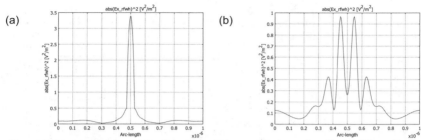

Figure 9. Line scans taken at the object (a) and image (b) planes described in Figure 7.

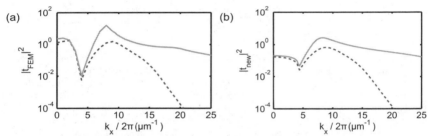

Figure 10. (a) FEM-derived transfer functions for single- (dashed) and triple-layer (solid) Ag superlenses with (b) the reflection-modified analytical transfer function shown for comparison.

DISCUSSION AND CONCLUSIONS

The two different modified transfer functions (t_{new} and t_{FEM}) are now directly comparable with the simple analytical transfer functions shown in Figure 2 (a). The three cases care compared directly in Figure 11, with the single-layer-superlens transfer functions shown in Figure 10 (a) and those for the triple-layer lens in Figure 10 (b). From this comparison it is seen that the position of the dominant resonant peak in the transfer function occurs at higher spatial frequencies for the modified transfer function (t_{new}) and the FEM-based transfer function (t_{FEM}), and that the high-spatial-frequency performance for these two cases is also very similar.

The major difference between t_{new} and t_{FEM} occurs for low spatial frequencies, where t_{FEM} is significantly higher than both t and t_{new}. We attribute this to the finite domain size that is required for the FEM simulations, as this introduces inaccuracies for determining Fourier components of the fields for spatial frequencies smaller than the inverse of the domain size ($1\ \mu m^{-1}$ in this case). In addition, non-ideal PMLs at the domain boundaries can introduce unwanted reflected fields that will tend to add to the true transmitted fields, causing the low-spatial-frequency transfer function that is extracted from the FEM simulations to be larger than the true value. Simulations with larger domain sizes could be used to reduce such errors however we note that this will compromise the ability to use a sufficiently large mesh density to obtain accurate results at high spatial frequencies. As these systems are designed for high-spatial-frequency imaging, trading off the accuracy at low spatial frequencies was considered to be the best choice for this initial study.

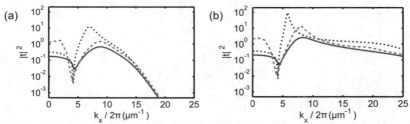

Figure 11. (a) Single-layer and (b) multi-layer silver superlens transfer functions. Modified analytical transfer functions, t_{new} (solid), FEM-derived transfer functions, t_{FEM} (dashed), and simple analytical transfer functions, t (dotted) are compared.

Finally, the improved analytical transfer function, t_{new} can be used to generate field profiles for representative imaging scenarios, for comparison with the field profiles that are generated using the simple analytical transfer function t. Examples are shown in Figure 12 for the dual-line-pair object of Figure 3 that is regularly used as a test of superlens performance [8]. It is seen that the damping of the resonances in the high-spatial-frequency region of the transfer function reduces the peak intensities in the images, although resolution is maintained. There is some evidence that the damping has reduced unwanted ringing effects that are often predicted [10] for these systems, although a more thorough investigation of this would be required before any significant conclusions could be drawn.

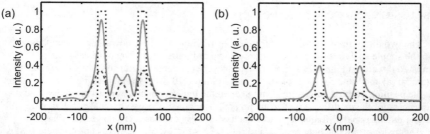

Figure 12. Image profiles from an ideal line-pair object (dotted) with 100-nm centre-to-centre spacing for both single-layer (dashed) and multi-layer (solid) superlenses. (a) Calculated using the simple analytical transfer function t, and (b) the modified transfer function t_{new}.

In conclusion, we have presented an improved method for determining the spatial-frequency transfer function of planar 'superlens' systems by taking into account interactions between the near-field-coupled object and lens. For an object that is represented by narrow apertures in an otherwise-continuous metal screen this is accounted for analytically using a simple multiple-reflection formula. We have shown that this gives a transfer function with high-spatial-frequency structure that agrees well with FEM full-field electromagnetic simulations. Improvements to the method will be required for other object classes, but this provides a useful starting point for the rapid evaluation of the influence of object / lens in coupling in near-field superlens imaging, a topic that has not been considered in any great detail to date.

ACKNOWLEDGMENTS

This work has been supported by the MacDiarmid Institute for Advanced Materials and Nanotechnology, and the Marsden Fund administered by the Royal Society of New Zealand. One of us (CPM) also acknowledges financial support from the Department of Electrical and Computer Engineering, University of Canterbury.

REFERENCES

1. Betzig, E., et al. (1991) 'Breaking the diffraction barrier - optical microscopy on a nanometric scale', *Science*, **251**, 1468-1470.
2. Alkaisi, M.M., Blaikie, R.J., McNab, S.J., Cheung, R. and Cumming, D.R.S. "Sub-diffraction-limited patterning using evanescent near field optical lithography". *Appl. Phys. Lett.* **75**, 3560-3562 (1999).
3. J. B. Pendry, "Negative refraction makes a perfect lens," *Phys. Rev. Lett.* **85**, 3966 (2000).
4. Melville, D.O.S. and Blaikie, R.J. (2005) 'Super-resolution imaging through a planar silver layer', *Opt. Express*, **13**, 2127-2134.
5. Fang, N., Lee, H., Sun, C., and Zhang, X. (2005) 'Sub-diffraction-limited optical imaging with a silver superlens', *Science*, **308**, 534-537.
6. C. Luo, S. G. Johnson, J. D. Joannopoulos, Phys. Rev. B **68** (2003) 045115.
7. D. R. Smith, D. Schurig, M. Rosenbluth, S. Schultz, S. A. Ramakrishna, J. B. Pendry, Appl. Phys. Lett. **82** (2003) 1506-1508.
8. S. A. Ramakrishna, Rep. Prog. Phys. **68** (2005) 449-521.
9. D. O. S. Melville, R. J. Blaikie, Physica B **394** (2007) 197-202.
10. Moore, C.M., Arnold, M.D., Bones, P.J. and Blaikie, R.J. "Image fidelity for single- and multi-layer silver superlenses", *J. Opt. Soc. Am. A* **25**, 911-918 (2008).
11. V. G. Veselago, "The electrodynamics of substances with simultaneously negative values of ε and μ," Soviet Physics Uspekhi, vol. 10, no. 4, pp. 509–514, 1968.
12. Johnson, P.B. and Christy, R.W. (1972) 'Optical-constants of noble-metals', *Phys. Rev. B*, Vol. 6, No. 12, pp. 4370-4379.
13. D. R. Lide, *The CRC handbook of chemistry and physics*, 88th ed. CRC Press, 2008.
14. COMSOL is a registered trademark of COMSOL AB, © 1997–2009.

Poster Session II

Mater. Res. Soc. Symp. Proc. Vol. 1182 © 2009 Materials Research Society   1182-EE13-02

Composites of silver nanoparticles and poly(vinylidene fluoride-trifluoroethylene) copolymer: Preparation as well as structural and electrical characterization

Tonino Greco and Michael Wegener

Functional Materials and Devices, Fraunhofer Institute for Applied Polymer Research (IAP), Geiselbergstraße 69, 14476 Potsdam, Germany

ABSTRACT

In this work, the influence of homogeneously dispersed silver nanoparticles in the ferroelectric copolymer poly(vinylidene fluoride-trifluoroethylene) (P(VDF-TrFE)) has been investigated. Metallic nanocomposite films based on silver particles embedded in a P(VDF-TrFE) matrix were prepared as 20 µm thin films by solvent casting and following in situ formation of silver colloids within the solid matrix by UV irradiation and thermal treatment. Within the 0–3 composites, the metallic-mass fraction was varied between 0.01 and 1 wt.%, which yielded films with different optical properties. In the prepared nanocomposite films the surface plasmon polariton resonance peak is observed with increasing extinction in the visible spectral range. From the infrared spectra and the measured thermal properties it is concluded that no significant degradation of the polymers occurs. Finally, electrical poling was performed in order to investigate the influence of the embedded silver particle on the polarization build up of the P(VDF-TrFE) matrix polymer. From initial experiments it is concluded that conductivity phenomena across the sample thickness are prevented and a dipolar polarization in the P(VDF-TrFE) is achieved.

INTRODUCTION

Nanocomposites consisting of polymers and metal nanoparticles are of great interest in regard to electronic and opto-electronic applications [1-3]. The preparation of such nanocomposites with homogeneously dispersed particles cannot be carried out normally by mixing the polymer and the desired isolated colloids due to strong agglomeration tendency of the metallic nanoparticles. Consequently, nanocomposites with colloids have been prepared by synthesis of the inorganic particles in situ, for instance in solution, and then mixed with the polymer solution. Extensive attention has been given to the study of the plasmonic properties of noble metal nanoparticles as a result of their potential application as waveguides, photonic circuits and sensors [4-6]. Surface plasmon polaritons are excited when electromagnetic radiation causes coherent oscillations of the conducting electrons of noble metal nanoparticles such as gold, silver or copper. The selective photon absorption and scattering of electromagnetic waves by the nanoparticles can be observed by conventional methods like UV-vis spectroscopy and reflection measurements. The surface plasmon resonance frequency is extremely sensitive to the size, shape, interparticle separation and the surrounding dielectric environment of the nanoparticles [7]. By rising the noble metal nanoparticle content in polymers, the absorption bandwidth can be tuned from the near ultraviolet (UV, 200-400 nm) across the visible (vis, 400-700 nm) to the near and far infrared (NIR, 700-3000 nm, IR, > 3 nm) as very recently shown [8]. The absorption of light and excitation of surface plasmons can lead to strongly enhanced electric fields within the metallic nanoparticles and the inter-nanoparticle gaps. The ability to extend plasmon resonance absorptions beyond visible wavelengths of 400-700 nm is important for various potential IR applications, including thermal imaging and infrared photo-detectors and pyroelectric sensors [9, 10].

In order to obtain multifunctional composites electro-active polymers (EAP) can be chosen as matrix materials. Suitable polymer candidates are electrets such as polyvinylidene fluoride (PVDF) [11] as well as its copolymers with trifluoroethylene (P(VDF-TrFE)) [12] or hexafluoropropylene (P(VDF-HFP)) [13]. These polymers show ferroelectric polarization accompanied with piezo- and pyroelectric properties and they were chosen as matrix materials for different kind of ceramic-polymer composites as exemplarily shown in References [14, 15]. Recently, also composites of PVDF with embedded metallic nanoparticles were studied regarding the kinetics of film preparation, particle dispersion and their resulting properties [16]. However, there only a few studies about the effects of surface plasmon resonances caused by metallic nanoparticles in ferroelectric polymers like PVDF or P(VDF-TrFE).

Here, we present a study of nanocomposites comprised of silver particles formed in situ in P(VDF-TrFE). The thermal as well as UV-vis and infrared properties of such nanocomposites are discussed in detail. Finally the composite films were metallized on both film surfaces and their poling behavior is studied by applying high electric fields and by analyzing the poling currents.

EXPERIMENT

P(VDF-TrFE) powder with a VDF/TrFE composition of 77/23 (Piézotech S.A., France), silver nitrate $AgNO_3$ (Aldrich, $\geq 98.5\%$) and dimethyl formamide (DMF, Fluka, $\geq 99.8\%$) were chosen for the composite preparation and not purified before use. The P(VDF-TrFE) powder as well as the $AgNO_3$ were separately dissolved in DMF. Both solutions were mixed using a magnetic stirrer. Composite films were prepared by solvent casting the mixture onto glass plates (5 cm × 5 cm) at a temperature of 50°C. The cast P(VDF-TrFE)-$AgNO_3$ composite films were first dried on a hot plate at a temperature 80°C for 1 hour, for 10 min at 165°C and further 2 hours at 120°C. Finally, the films were cooled down to room temperature. Afterwards, the dried films were irradiated by UV-C light ($\lambda = 250$ nm) with an exposure time of 45 min. The thickness of the Ag-P(VDF-TrFE)-nanocomposite films was adjusted to 20 µm. The prepared composite films exhibit silver nanopartciles from 0.01 to 1 wt.%. Basis of this calculation is the assumption that the decomposition of silver nitrate to silver occurs completely. The nanocomposite films were removed from the glass slides by emerging them into distilled water for 10 minutes and detaching them with a razor blade. In addition pure P(VDF-TrFE) films were prepared as reference by using the above described heating procedure for the solvent evaporation.

The excitation of surface plasmon resonance of silver nanoparticles in the thin polymer films were characterized by an UV-vis-NIR spectrophotometer (Perkin Elmer Lambda 950). FTIR spectra were recorded using an IR spectrophotometer (Thermo Nicolet Nexus 470) with an ATR (attenuated total reflection) device. Thermal properties were investigated by means of differential-scanning calorimetry (DSC) and carried out at a rate of 10 K/min. Electrical poling was performed in direct contact at room temperature. During direct-contact poling combinations of bipolar and unipolar poling cycles were applied to the samples via a computer controlled bipolar high-voltage device (TREK 610C). The poling current was measured in situ by means of an electrometer (Agilent 3458a) operated in the current mode. Current contributions from the charging of the sample capacitance and the conductivity were subtracted from the measured poling currents according to the procedure described in [17]. The advantage of the here used technique is that current contributions from charging of the sample capacitance and from the conductivity are measured directly and no theoretical assumptions of these values are necessary. However, the disadvantage is that the calculation of the polarization yields two half hysteresis cycles, one for negative and one for positive poling fields, instead of a complete hysteresis loop, which have to be adjusted by taking into account the maximum of the poling-current peaks. All measurements were performed on both-side aluminum metallized samples.

DISCUSSION

Silver nitrate is known to decompose by heat treatment or by UV-light exposure by following the below given reaction.

$$AgNO_3 \xrightarrow[h\nu]{\Delta} Ag + NO_2 + \tfrac{1}{2} O_2$$

Most of silver cations in P(VDF-TrFE) are reduced to metallic silver after thermal treatment at 165°C and irradiation with UV light. The high dispersion of $AgNO_3$ in the polymer matrix favors the desired in situ reduction to silver colloids. The degree of the reduction can be easily measured in infrared spectra by monitoring the intensity of the characteristic bands due to nitrate at $801\ cm^{-1}$ and $733\ cm^{-1}$. For all $AgNO_3$ contents the IR spectra of Ag-P(VDF-TrFE) composites are similar to those of pure P(VDF-TrFE) films indicating that the transformation of the silver precursor to Ag nanoparticles is complete. Thus, the thermal treatment at 165°C and the following UV-light exposure for 45 minutes seem to be sufficient for the decomposition of low $AgNO_3$ contents. In addition, as shown in Figure 1 the infrared spectra of different kinds of nanocomposites demonstrate that the preparation procedures do not lead to a decomposition of the polymeric matrix.

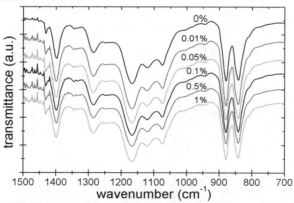

Figure 1: ATR-IR spectra of free-standing Ag-P(VDF-TrFE)-nanocomposite films.

This conclusion is also supported by the thermal properties recorded during DSC measurements. As shown in Figure 2, the measured melting temperatures T_M of the composites are comparable to the melting temperature of P(VDF-TrFE). Furthermore, the peak in the heat flow describing the Curie transition T_C in the P(VDF-TrFE) copolymer is for all Ag-P(VDF-TrFE) compositions at the same temperature of 125°C as expected for this kind of PVDF copolymer.

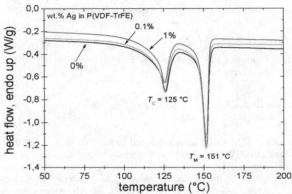

Figure 2: DSC spectra of synthesized Ag-P(VDF-TrFE)-nanocomposite films.

For all kind of nanocomposite films typical plasmon-resonance absorptions are found in the UV-vis spectra. In Figure 3 the spectra recorded on composite samples on glass substrates are compared. In conclusion, the particle sizes increase slightly with the increase of the silver nitrate content in order to get a silver content of about 1 wt.% resulting in a slight shift of the plasmon-resonance peak as shown in Figure 3.

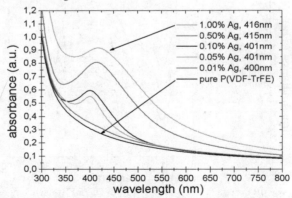

Figure 3: UV-vis-extinction spectra of Ag-P(VDF-TrFE)-nanocomposite films.

During poling experiments with applied electric fields up to 90 MV/m a polarization of about 75 mC/m² and a coercive field of approximately 45 MV/m is found in pure P(VDF-TrFE) films prepared by means of the above discussed procedures, solvents and heating routines. A quite similar switching in the measured poling currents is observed in all Ag-P(VDF-TrFE) composites with silver contents of up to 1 wt.%. Based on the found saturation in the poling currents and due to the relatively short time of the applied electric field a dipolar polarization within the composites can be assumed. As an example, the measured poling current and the calculated polarization – electric field hysteresis of an Ag-P(VDF-TrFE) composites with 1 wt.% silver are shown in Figure 4.

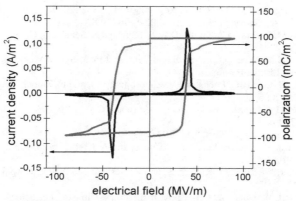

Figure 4: Ag-P(VDF-TrFE)-nanocomposite with 1wt.% Ag: Measured poling current and calculated polarization as function of the applied electric field.

As found in these first experiments, the determined coercive fields are within the measurement error of the film thicknesses, however, the calculated polarization increases very slightly with increasing amount of silver particles. Furthermore, it is observed that both calculated polarization – electric field dependencies are not exactly symmetric as for instance in pure P(VDF-TrFE). Possible space-charge effects, influences of the metal-organic interface as well as the interaction of the metallic particles and the polymer matrix during the poling process and the resulting internal electric fields are under further investigation.

CONCLUSIONS

In summary, it is demonstrated that noble-metal nanoparticles are formed in a polymer matrix which is able to show ferroelectric properties after further processing steps. The developed procedure includes good controllable processing steps such as thin-film casting as well as thermal and photo-induced decomposition of silver precursors. By adjusting the silver content it is possible to gain thin films of polymer nanocomposites with effective and long-range light absorption in the UV and visible spectral range. A significant dipolar polarization is achieved in all nanocomposites after suitable electrical poling. The synthesized thin films are strong candidates for multifunctional composites showing the discussed optical absorption properties as well as the ferroelectric polarization and thus following electro-active properties such as piezo- and pyroelectricity.

REFERENCES

1.   E. Hutter and J.H. Fendler, Adv. Mater. 16, 1685-1706 (2004).
2.   S.A. Maier, M.L. Brongersma, P.G. Kik, S. Meltzer, A.A.G. Requicha, and H.A. Atwater, Adv. Mater. 13, 1501-1505 (2001).
3.   S. Link and M.A. El-Sayed, J. Phys. Chem. B 103, 8410-8426 (1999).
4.   S.A. Maier and H.A. Atwater, J. Appl. Phys. 98, 011101 (2005).
5.   N.L. Rosi and C.A. Mirkin, Chem. Rev. 105, 1547-1562 (2005).
6.   S. Nie and S.R. Emory, Science 275, 1102-1106 (1997).
7.   J.J. Mock, M. Barbic, D.R. Smith, D.A. Schultz and S. Schultz, J. Chem. Phys. 116, 6755-6759 (2002).
8.   T. Greco and M. Wegener, Proceeding, 5th Intern. Mater. Symp. Materiais 2009, 05.-08.04.2009, Lisboa, Portugal.
9.   H. Takele, H. Greve, C. Pochstein, V. Zaporojtchenko, and F. Faupel, Nanotechn., 17, 3499-3505 (2006).
10.   A. Biswas, H. Eilers, J.F. Hidden, O.C. Aktas, and C.V.S. Kiran, Appl. Phys. Lett. 88, 013103 (2006).
11.   H. Kawai, Jpn. J. Appl. Phys. 8, 975-977 (1969).
12.   T. Furukawa, M. Date, E. Fukada, Y. Tajitsu, and A. Chiba, Jpn. J. Appl. Phys. 19, L109-L112 (1980).
13.   M. Wegener, W. Künstler, K. Richter, and R. Gerhard-Multhaupt, J. Appl. Phys. 92, 7442-7447 (2002).
14.   K.L. Ng, H.L.W. Chan and C.L. Choy, IEEE Trans. Ultrason. Ferroelectr. Freq. Contr. 47, 1308-1315 (2000).
15.   M. Wegener and K. Arlt, J. Phys. D: Appl. Phys. 41, 165409 (2008).
16.   J. Compton, D. Kranbuehl, G. Martin, E. Espuche, and L. David; Makromol. Symp. 247, 182-189 (2007).
17.   M. Wegener, Rev. Sci. Instr. 79, 106103 (2008).

Mater. Res. Soc. Symp. Proc. Vol. 1182 © 2009 Materials Research Society   1182-EE13-03

Strong Exciton Polariton Dispersion in Multimode GaN Microcavity

M. C. Liu[2], Y.-J. Cheng[1,2], S.-H. Hsu[1], H. C. Kuo[2], T. C. Lu[2], and S. C. Wang[2]
[1]Research Center for Applied Sciences, Academia Sinica, Taipei 11529, Taiwan
[2]Department of Photonics and Institute of Electro-Optical Engineering, National Chiao Tung University, 1001 TA Hsueh Rd., Hsinchu 300, Taiwan

ABSTRACT

We report the experimental observation of a very strong cavity polariton dispersion in a multi-axial mode GaN microcavity. The linewidth of photoluminescent (PL) spectrum covers a few cavity axial modes. The resonant photoluminescent peaks have a strong dispersion. The frequency spacing between adjacent peaks decreases by almost a factor of five from 470nm to 370nm. The strong dispersion can be well described by cavity polariton dispersion, but not by the dispersion of the refractive index of GaN. The measured exciton-photon interaction constant is 260 meV. It is an order of magnitude higher than the typically reported values for GaN microcavities

INTRODUCTION

The strong interaction between excitons and photons in semiconductor microcavities has generated great research activities in the past few years. In the strong interaction regime, excitons and cavity photons form bosonic quasi particle states, the so called cavity polaritons, which can condensate to the ground state through polariton scattering processes [1-5]. The condensation of cavity polaritons can lead to polariton lasing and spontaneous coherent radiation is emitted. The experiments have been carried out mostly in planar semiconductor microcavities. The cavity polariton dispersion curve and/or Rabi splitting spectrum are often used as an indication for strong exciton-photon interaction. To promote strong exciton-photon interaction, a microcavity with short cavity length is commonly used. The microcavities fabricated for polariton experiments have exclusively used cavity length of just a few integer multiples of half-wavelengths, where there is only one cavity mode in interaction with excitons. A long cavity has never been used for strong exciton-photon interaction experiments.

Here we report the observation of a strong exciton multi-mode photon interaction in a GaN microcavity. The cavity length is so long that there are multi-cavity axial modes simultaneously in interaction with excitons. The observed interaction constant is much higher than those reported for short microcavities. The cavity length of our device is a few tens of GaN transition wavelength, much longer than the typical length used in cavity polariton experiments. The estimated cavity axial mode spacing, based on the physical cavity length and the nominal GaN index of refraction, is smaller than the as grown GaN photoluminescent (PL) linewidth. Normally, one would expect to see multiple axial mode peaks with equal frequency spacing in PL spectrum. However, the PL measurement has shown a rather strong dispersion around GaN transition frequency. The axial mode frequency spacing decreases significantly with increasing frequency from below GaN bandgap energy. This dispersion can be well described by the cavity polariton dispersion equation. An ellipsometer measurement of the index of refraction of a similar GaN sample without cavity confirms that this is not due to the refractive index dispersion. The exciton-photon interaction constant derived from the measured polariton dispersion is as high as 260 meV, the largest value reported so far to the best of our knowledge. It is an order of magnitude larger than the best values reported for semiconductor microcavities, where cavity lengths are just a few half-wavelengths [6-8].

EXPERIMENT

The GaN/InGaN microcavity device was fabricated by a standard epitaxial growth, followed by dielectric coating, laser lift off, and another dielectric coating to finally form a surface emitting microcavity, as shown in Fig. 1. The device was grown on a (0001)-oriented sapphire substrate by metalorganic chemical vapor deposition(MOCVD). The layer structures are: 10nm nucleation layer, a 4 μm bulk GaN layer, GaN/InGaN multi quantum wells (MQW), followed by a final 200nm GaN cap layer. The InGaN MQWs consist of 10 pairs of 5 nm GaN barrier and 3 nm of $In_{0.1}Ga_{0.9}N$ well. The PL emission peak of the fabricated MQW was at 420 nm. A 6 pairs of SiO_2/TiO_2 dielectric distributed Bragg reflector(DBR) was deposited on the top surface. A silica substrate was epoxied onto the DBR surface. A pulsed excimer laser was then focused through sapphire substrate onto sapphire GaN interface to remove sapphire substrate by thermal ablation. The GaN wafer was transfered to the silica substrate after this laser lift-off process. The exposed GaN surface was mechanically polished to mirror quality, followed by a SiO_2/Ta_2O_5 DBR dielectric coating. The mechanical polish of GaN surface only removed a few tens of nm of GaN. The final finished Fabry-Perot cavity formed by these two DBR mirrors has a cavity length of 4.2 μm. The DBRs were designed to have $>99\%$ reflectivity from 400 nm to 450 nm to cover QW emission wavelength. The cavity Q value was estimated to be around 900 based on a low pump power PL spectrum. The DBR reflectivity has a roll off from 90\% reflectivity at 383 nm to the first reflectivity minimum at 368 nm. The 4 μm bulk GaN layer has a PL peak at 365 nm with a linewidth of 7 nm. The Lorentzian roll off of GaN PL spectrum still has 7% of its peak value at 380 nm. The roll off of DBR reflectivity and that of the Lorentzian tail of GaN PL spectrum are in opposite directions but still have a reasonable overlap between 370 nm and 380 nm. Normally, this is not an optimized cavity reflectivity condition for investigating exciton-photon interaction. Nevertheless, axial mode peaks in the overlap region of the roll offs between reflectivity and GaN PL could still be clearly identified and an unusual strong resonant mode dispersion was observed.

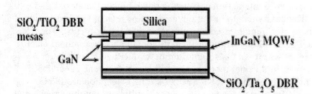

Fig.1 GaN surface emitting microcavity.

The fabricated microcavity was pumped by a CW HeCd laser through the SiO_2/Ta_2O_5 DBR mirror at room temperature. The laser was focused by a 15X objective with spot size about 1μm. The PL emission was collected by the same objective and coupled into the fiber input of a spectrometer. The PL spectra at different pump power levels are shown in Fig. 2(a) along with a zoom in Fig. 2(b) around the GaN wavelength region. The GaN and InGaN QW transitions, 365 nm and 420 nm, were both excited. Multimode peaks can be identified from 370 nm all the way to 470 nm. Multiple Lorentzian profiles are used to fit the spectral peaks from 360 nm to 390 nm to identify their exact locations. Two typical fitting results are shown in Fig. 2(c) and (d) for high and low pump power levels. The sums of the fitted Lorentzian profiles are also displayed and both show fairly good fits. The peaks shorter than 370 nm becomes less obvious as pump power decreases and their locations obtained from curve fitting are less certain in particular for the lowest two pump power levels. For those peaks can be clearly identified, the fitted peak positions do not have noticeable changes within the range of different pump power levels. The fitted linewidths for the center most obvious three peaks slightly decrease as pump power increases. For example, the linewidth of 372.4nm peak decreases from 1.5nm to 1.26nm as pump power increases from at 5.7mW to 12.3mW. The PL spectrum are randomly polarized. The mode spacing decreases by almost a factor of five from 470 nm to 370 nm. We attribute this large dispersion to strong exciton-photon interaction.

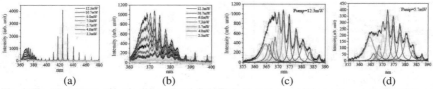

(a)   (b)   (c)   (d)

Fig. 2 The PL spectra of optically pumped GaN/InGaN surface emitting microcavity. (a) PL spectra at various pump power levels. (b) A zoom in spectrum around GaN transition wavelength region. (c) (d) Typical multiple Lorentzian profile fitting along with the sums of the fitted profiles superimposed on the measured PL spectra.

DISCUSSION

In the strong interaction regime, exciton and photon form two coupled cavity polariton states. The two polariton states, upper and lower branch, have an unique anti-crossing dispersion characteristic. The observed resonant frequencies versus cavity axial mode wave numbers are shown with red square legends in Fig. 3. The observed peaks are fitted by the lower branch cavity polariton dispersion equation,

$$\omega_{pol,n} = (\Omega + \omega_n)/2 - \sqrt{(\Omega - \omega_n)^2 + 4g^2}/2$$

Eq. (1)

where Ω is the exciton frequency, ω_n is the photon frequency of the nth cavity mode, $\omega_{pol,n}$ is the corresponding polariton frequency, and g is the exciton photon interaction constant. The cavity mode frequency in the form $\omega_n = a + n\,\Delta\omega$ is assumed in the fitting, where a and mode frequency spacing $\Delta\omega$ are fitting variables with mode index integer n starting from 1 for the lowest observed resonant frequency. The blue line is the fitted curve and it shows an excellent fit with the fitting parameters, $h\Omega$=3.50 eV, hg=0.26 eV, ha=2.601 eV, and mode spacing h $\Delta\omega$ = 0.0635 eV. The exciton energy $h\Omega$ =3.50 eV from fitting is reasonably close to the 3.45 eV value cited in literatures. The justification of the fitting using polariton dispersion equation is further assured by the fact that a/ $\Delta\omega$ =40.96 is only off from the closest integer by 4%. Ideally, cavity mode frequency ω_n = a + n $\Delta\omega$ should be an exact integer multiple of $\Delta\omega$, i.e. a should be an integer multiple of $\Delta\omega$. The offset of merely 4% of $\Delta\omega$ can be regarded as an excellent validation of the fitting model. The exciton and photon energies versus wave number k obtained from the curve fitting are also shown with two straight lines where the black triangle legends are the corresponding cavity modes coupled to excitons to form cavity polaritons. The interaction constant of hg=260 meV is the highest value reported so far to the best of our knowledge. It is an order of magnitude larger than the recently reported state of the art values obtained from Rabi splitting measurements in III-nitride based devices, where Rabi splittings 2hg of 50 meV and 56 meV were reported for 3λ/2 and 3λ cavity respectively [6-8].

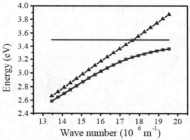

Fig. 3. Multimode cavity polariton dispersion. The red square legends are the observed resonant energies plotted versus equally spaced wave numbers. The blue line is the fitting curve based on cavity polariton dispersion equation. The two straight lines are the exciton and cavity mode energies obtained from fitting parameters. The black triangle legends are the cavity photon modes.

CONCLUSIONS

In summary, we have observed a very strong polariton dispersion in a multimode GaN surface emitting microcavity at room temperature. There are multiple photon modes simultaneously in interaction with exciton. The frequency spacing between adjacent PL peaks decreases by almost a factor of five from 470 nm to 370 nm. The dispersion in PL peaks can be described very well by the lower branch cavity polariton dispersion equation. It is shown that the dispersion is not due to the intrinsic material dispersion. The fitting gives an exciton-photon interaction constant of 260 meV, the highest value reported so far.

The authors would like to acknowledge the support by the National Science Council of Taiwan under Contract No. NSC 97-2112-M-001-027-MY3.

REFERENCES

1. A. Imamoglu, R. J. Ram, S. Pau, and Y. Yamamoto, Phys. Rev. A **53**, 4250 (1996).
2. R. Butte, G. Delalleau, A. I. Tartakovskii, M. S. Skolnick, V. N. Astratov, J. J. Baumberg, G.Malpuech, A. D. Carlo, A. V. Kavokin, and J. S. Roberts, Phys. Rev. B **65**, 205310 (2002).
3. G. Malpuech, A. D. Carlo, A. Kavokin, J. J. Baumberg, M. Zamfirescu, and P. Lugli, Appl. Phys. Lett. **81**, 412 (2002).
4. S. Christopoulos, G. Baldassarri Hoger von Hogersthal, A. J. D. Grundy, P. G. Lagoudakis, A. V. Kavokin, J. J. Baumberg, G. Christmann, R. Butte, E. Feltin, J.-F. Carlin, et al., Phys. Rev. Lett. **98**, 126405 (2007).
5. T. Tawara, H. Gotoh, T. Akasaka, N. Kobayashi, and T. Saitoh, Phys. Rev. Lett. **92**, 256402 (2004).
6. G. Christmann, R. Butte, E. Feltin, A. Mouti, P. A. Stadelmann, A. Castiglia, J.-F. Carlin, and N. Grandjean, Phys. Rev. B **77**, 085310 (2008).
7. E. Feltin, G. Christmann, R. Butte, J.-F. Carlin, M. Mosca, and N. Grandjean, Appl. Phys. Lett. **89**, 071107 (2006).
8. G. Christmann, R. Butt´e, E. Feltin, J.-F. Carlin, and N. Grandjean, Appl. Phys. Lett. **93**, 051102 (2008).

Mater. Res. Soc. Symp. Proc. Vol. 1182 © 2009 Materials Research Society   1182-EE13-05

Effective Excitation of Superfocusing Surface Plasmons Using Phase Controlled Waveguide Modes

Kazuhiro Yamamoto[1], Kazuyoshi Kurihara[1], Junichi Takahara[2] and Akira Otomo[1]

[1] Kobe Advanced ICT Research Center, National Institute of Information and Communications Technology (NICT), 588-2 Iwaoka, Nishi-ku, Kobe 651-2492, Japan

[2] Graduate School of Engineering Science, Osaka University, 1-3, Machikaneyama-cho, Toyonaka, Osaka, 560-8531, Japan

ABSTRACT

We propose a more practical method of realizing the superfocusing modes based on waveguide structures, and present a numerical analysis of these structures using finite-difference time-domain (FDTD) simulations. For metallic wedged structures coupled to dielectric waveguides, we investigate a method of controlling superfocusing by changing the phase of waveguide modes.

INTRODUCTION

Recently, surface plasmons have attracted attention as information carriers for novel nano optical systems. As surface plasmons are localized electromagnetic fields on metal-dielectric interfaces or metallic nanostructures, one can construct integrated optical systems that overcome the diffraction limit of light. In this context, a special electromagnetic mode, called "superfocusing modes," is garnering the attention of researchers, owing to the high field concentration effect due to the increase of the wavenumber of surface plasmons [1-5]. These modes are usually excited in tapered metallic structures, like cones and wedges, with specific excitation polarizations. For instance, in the case of metallic cones, the polarization of excitation of the electric field must be radial. Superfocusing of surface plasmons in tapered simple metallic structures has been studied theoretically [1-5]. But there are no effective excitation methods and realistic structures of such superfocusing surface plasmons for experiments. In this presentation, we propose a novel method of exciting superfocusing of surface plasmons in wedged metallic structures based on the phase controlled gap waveguide mode, and present a numerical analysis of these structures using finite-difference time-domain (FDTD) simulations [6].

NUMERICAL CALCULATION

The schematic structure is shown in Figure 1. This structure consists of two adjacent gap (metal-dielectric-metal) waveguides that couple dielectric waveguides and the wedged metallic structure between the two gap waveguides. When the phases of propagation modes of two waveguides are out of phase, we realized the two gap plasmon modes with opposite polarization in the gap waveguide section [7, 8]. These gap modes propagate to the wedged section and are converted to surface plasmons at the interface with appropriate polarization for superfocusing

modes. As a result, electric field concentration occurs at the apex of the metal wedge. In contrast, if the phases of the modes are in phase, the polarization at the interface of the wedge is not appropriate and superfocusing does not occur.

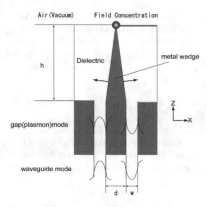

Figure 1. Schematic picture of superfocusing of surface plasmons in wedged metal structure with dielectric waveguides.

To verify the validity of the above scheme, we numerically simulate the propagation behavior of surface plasmons in such structure by the 2D-FDTD method. Numerical results were calculated using the RSoft *FullWAVE* simulation tool with a 2.5 nm grid under the Courant stability bound [6].

RESULTS AND DISCUSSION

In the case where the metal is gold and the dielectric is silicon, when the structure is excited at a telecommunication wavelength of 1550 nm we observed the increase of the wavenumber along the tapers of wedge and the enhancement of the longitudinal electric field at the apex. Figure 2 shows spatial distribution of the longitudinal (z) electric field when the phases of propagation modes of two waveguides are out of phase. For typical geometric parameters (center metal thickness is d=250 nm, gap width is w=300 nm and wedge length is h=1000 nm), the full width of half of the maximum of the electromagnetic energy density distribution is 5 nm and the square of longitudinal electric field increased by a factor of 10 at the apex.

Figure 3 (a) shows the electromagnetic energy density distribution along the surface of the taper under the out of phase condition. It can be clearly observed that the wavenumber and the electromagnetic energy density of surface plasmon increase toward the apex.

On the one hand, Figure 3(b) shows the result when the structure is excited with two in phase waveguide modes. In this case, the features of superfocusing are not observed and electromagnetic energy density decreases along taper due to ohmic loss.

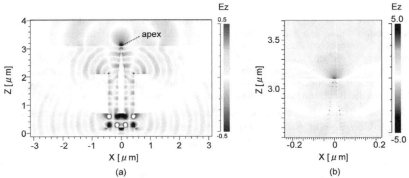

Figure 2. Spatial distribution of longitudinal (z) component of electric field (out of phase). (a) in the whole structure (b) near the apex.

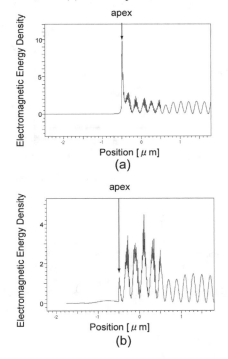

Figure 3. Electromagnetic energy density distribution along the metal taper. (Phase of two waveguide mode:(a) out of phase, (b) in phase.)

With the above results, we can control the superfocusing using the phase difference of two waveguide modes. These phase difference can be realized easily with changing waveguide lengths or the electro-optical modulation technique.

Superfocusing of surface plasmons in wedged structures depends on the taper angle. Figure 4 shows taper length (equivalently taper angle) dependence of the longitudinal field enhancement at the apex. While the theory predicted superfocusing would occur efficiently because of the smaller taper angle, the result shows enhancement is peaked at the appropriate taper length (angle) [5]. In this case, electric field enhancement peaked at h = 1μm. This is because the theory does not include the loss of metal. As our simulations include the losses, the electric field enhancement decreases when the taper length is longer (taper angle is smaller).

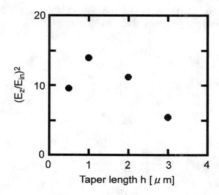

Figure 4. Taper length dependence of the longitudinal electric field enhancement at apex.

CONCLUSIONS

In this paper, we proposed a novel wedged metal structure with phase-controllable waveguides and an excitation method for superfocusing of surface plasmons. We conducted numerical simulations of these structures and methods and observed superfocusing features. By modulating the phase of waveguide modes, fast changing of local electric field can be realized. This scheme can be applied to waveguide-based nano optical control devices.

REFERENCES

1. Kh. V. Nerkararyan, Phys. Lett. A, **237**, 103-105 (1997).
2. A. J. Babadjanyan, N. L. Margaryan, and Kh. V. Narkararyan, J. Appl. Phys., **87**, 3785-3788 (2000).
3. M. I. Stockman, Phys. Rev. Lett., **93**, 137404-1-4 (2004).
4. K. Kurihara, A. Otomo, A. Syouji, J. Takahara, K. Suzuki, and S. Yokoyama, J. Phys. A: Math. Theor., **40**, 12479-12503 (2007).
5. K. Kurihara, K. Yamamoto, J. Takahara, and A. Otomo, J. Phys. A:Math. Theor., **41**, 295401(48pp) (2008).
6. A. Taflove and S. C. Hagness, *Computational Electrodynamics. The Finite-Difference Time-Domain Method*, 2nd ed., Artech House, Boston, MA (2000).
7. E. N. Economou, Phys. Rev, **182**, 539-554 (1969).
8. J. Takahara and F. Kusunoki, IEICE Tran. Electron., **E90-C**, 87-94 (2007).

Mater. Res. Soc. Symp. Proc. Vol. 1182 © 2009 Materials Research Society   1182-EE13-07

Fidelity of Holographic Lithography for Fabrication of 3D SU-8 Photonic Structures and How to Minimize Distortion by Optical Design

Xuelian Zhu, Yongan Xu, Shih-Chieh Cheng, and Shu Yang
Department of Materials Science and Engineering, 3231 Walnut Street, Philadelphia, PA 19104

ABSTRACT

Theoretical analysis can impart great benefits on the rationale design of 3D photonic structures by revealing the underlying mechanisms of structural distortion during each processing step. In this report, we quantitatively study the distortion of a three-term diamond-like structure fabricated in SU-8 polymer by four-beam interference lithography, which can be attributed to refraction at the air-film interface, and resist film shrinkage during lithographic process. In study of photonic bandgap (PBG) properties of Si photonic crystals templated by the SU-8 structures, we find that the distortion has degraded the quality of PBGs. Furthermore, we theoretically design new optical setups to fabricate three-term diamond-like structure with minimal deformation. Instead of single exposure of four beams, we use triple exposure of two beams, one from the central beam and the other from the side beam each time. A set of new linear polarization vectors is suggested to enhance the contrast between the minimal and maximal intensities of interference pattern.

INTRODUCTION

Photonic crystals (PCs)[1,2] with periodic dielectric microstructures are of interest for numerous applications in optical integrated circuits.[3] To realize three-dimensional (3D) photonic crystals with complete photonic bandgaps (PBGs), various methods have been developed to fabricate high-quality 3D photonic structures.[4-9] Among them, holographic lithography (HL) holds promise as an efficient and flexible technique for creating a wide range of defect-free 3D microstructures. During exposure, the multiple-beam interference intensity profile is transferred to a thick photoresist film, followed by post-exposure bake and development to create the microporous structures. The shape of the resulting structure is determined by the isodose surface of the lithographic threshold value, which can be described by the corresponding level surface.[10] In the level-set approach, the surface of a porous dielectric structure is represented by a scalar-valued function F, which satisfies $F(x, y, z) = t$, where t is a constant to control the volume fraction.

While many 3D photonic structures have been proposed theoretically, in HL experiments, the influence of refraction at the air/film interface cannot be ignored when the beams travel from the air into the resist film. To avoid this problem, many workers have applied prisms and/or index matching liquids[11-13] to precompensate the effects of refraction. However, the availability of specific prisms and index-matching liquids for many of structures and polymer systems could limit the broader application of HL for a wider range of 3D photonic structures. In addition, it is not uncommon to expect shrinkage of the photoresists during photoexposure and development, especially in the case of negative-tone resists, which could further distort the fabricated structures from the original design, resulting in decreased quality of PBGs.[14] Therefore, it is important to reveal the underlying mechanisms of structural distortion during each processing step and their impact to photonic bandgap properties. On the basis of the knowledge, we could exploit novel designs of optical systems to experimentally realize crystals with desirable structures and PBGs. In this report, the three-term diamond-like SU-8 structure fabricated by four-beam interference lithography (IL) with umbrella configuration was chosen as the model to discuss.

In order to precompensate the unidirectional shrinkage, one has to design proper beam geometry such that the shrinkage of photoresist will be balanced by a spatially elongated interference intensity profile, leading to a photonic crystal template with desired structural symmetry. Here, we theoretically design a new optical setup, two-beam triple exposure, to fabricate three-term diamond-like structure with minimal deformation. A set of new linear polarization vectors has been proposed to enhance the intensity contrast of interference pattern.

THEORETICAL ANALYSIS

The influence of refraction on wave vectors and polarization parameters for an arbitrary light beam can be quantitatively studied by Snell's law and Fresnel's equations.[15] The original parameters in air and the calculated new wave vectors and polarization vectors in the SU-8 for the four-beam umbrella HL were summarized in Table 1.

In our experiment, the visible laser source was split into four beams, where the central beam was right-circularly polarized and perpendicular to the surface of the photoresist SU-8 ($n_r \sim 1.60$ at $\lambda = 532$ nm), and the other three were linearly polarized and oblique at 39° relative to the central one. The circular polarization of the central beam distributes equal intensity to the surrounding beams and has been widely used in the fabrication of 3D PCs by HL.[16,17] The intensity ratio between the central beam and the side beams was 2:1:1:1. It is noted that no prism or index matching liquid was used in our optical setup, thus, the incident beams reached the surface of SU-8 film directly.

Table 1. Wave vectors and polarization vectors in air, SU-8 using IL with visible light ($\lambda = 532$ nm). Reprinted with permission from Ref. 14 Copyright © 2007 Optical Society of America.

Medium	Air	SU-8
Refractive index	1.0	1.6
Wave vectors	$\vec{k}_0 = \dfrac{2\pi}{\lambda}[\dfrac{1}{\sqrt{3}},\dfrac{1}{\sqrt{3}},\dfrac{1}{\sqrt{3}}];$	$\vec{k}'_0 = \dfrac{2\pi*1.6}{\lambda}[\dfrac{1}{\sqrt{3}},\dfrac{1}{\sqrt{3}},\dfrac{1}{\sqrt{3}}];$
	$\vec{k}_1 = \dfrac{2\pi}{\lambda}[\dfrac{5}{3\sqrt{3}},\dfrac{1}{3\sqrt{3}},\dfrac{1}{3\sqrt{3}}];$	$\vec{k}'_1 = \dfrac{2\pi*1.6}{\lambda}[0.852,0.371,0.371];$
	$\vec{k}_2 = \dfrac{2\pi}{\lambda}[\dfrac{1}{3\sqrt{3}},\dfrac{5}{3\sqrt{3}},\dfrac{1}{3\sqrt{3}}];$	$\vec{k}'_2 = \dfrac{2\pi*1.6}{\lambda}[0.371,0.852,0.371];$
	$\vec{k}_3 = \dfrac{2\pi}{\lambda}[\dfrac{1}{3\sqrt{3}},\dfrac{1}{3\sqrt{3}},\dfrac{5}{3\sqrt{3}}].$	$\vec{k}'_3 = \dfrac{2\pi*1.6}{\lambda}[0.371,0.371,0.852].$
Polarization vectors	$\vec{E}_0 = \sqrt{2}$ [circular polarization];	$\vec{E}'_0 = 1.088$ [circular polarization];
	$\vec{E}_1 = \pm[-0.25,0.345,0.905];$	$\vec{E}'_1 = \pm0.715\,[-0.484,0.286,0.827];$
	$\vec{E}_2 = \pm[0.905,-0.25,0.345];$	$\vec{E}'_2 = \pm0.715\,[0.827,-0.484,0.286];$
	$\vec{E}_3 = \pm[0.345,0.905,-0.25].$	$\vec{E}'_3 = \pm0.715\,[0.286,0.827,-0.484].$

RESULTS AND DISCUSSION

Refraction effect on both wave vectors and polarization parameters

From the beam parameters in air and SU-8, we calculated the corresponding reciprocal and real lattices (see Table 2 for respective lattice parameters).[18] The interference pattern in air was designed to have f.c.c translational symmetry. However, after considering the influence of refraction, the lattice was stretched along the [111] direction and became rhombohedral in SU-8 while there was no lattice distortion in the (111) plane.

Table 2. Lattice parameters for the interference patterns in air and in SU-8. ($c = 2\pi/d$ and $d = 1.38$ μm.) Reprinted with permission from Ref. 14 Copyright © 2007 Optical Society of America.

Medium	Air	SU-8 (considering refraction effect only)	SU-8 (considering both refraction and film shrinkage)
Plot of real lattices			
Translational symmetry	Fcc	Rhombohedral	Rhombohedral
Distance between the nearest lattice points in (111) plane (μm)	0.98	0.98	0.98
Distance between the adjacent lattice planes in [111] direction h_{111} (μm)	0.80	1.38	0.81
Space groups of the corresponding level surfaces	No. 166 ($R\bar{3}m$)	No.155 (R32)	No.155 (R32)

The change of lattice parameters due to refraction in turn alters the level surface that describes the structural symmetry. Given the central beam is right-circularly polarized, the level surfaces in air and in SU-8 were derived from the interference intensity profile as the following:

$$F_{air}(\vec{r}) = \cos[\frac{2\pi}{d}(-x+y+z)] + \cos[\frac{2\pi}{d}(x-y+z)] + \cos[\frac{2\pi}{d}(x+y-z)] \qquad (1)$$

and

$$F_{SU8}(\vec{r}) = \cos[\frac{2\pi}{d}(-1.140x+0.860y+0.860z)] + \cos[\frac{2\pi}{d}(0.860x-1.140y+0.860z)]$$
$$+ \cos[\frac{2\pi}{d}(0.860x+0.860y-1.140z)] + 0.301 \cos[\frac{2\pi}{d}\cdot 2(x-y)-\frac{\pi}{3}] \qquad (2)$$
$$+ 0.301 \cos[\frac{2\pi}{d}\cdot 2(y-z)-\frac{\pi}{3}] + 0.301 \cos[\frac{2\pi}{d}\cdot 2(z-x)-\frac{\pi}{3}]$$

where $d = 3\sqrt{3}\lambda/2 = 1.38$ μm. The structure is defined as

$$F(\vec{r}) > t \text{ for dielectric, and } F(\vec{r}) < t \text{ for air}$$

where t is a constant to control the filling volume fraction.

The level surfaces in air describes a three-term diamond-like structure in air, which belongs to space group No.166 ($R\bar{3}m$)[19,20] with inversed symmetry, whereas the symmetry of the level surface in SU-8 is No.155 ($R32$),[19,20] which has no inversion symmetry. At the same time, in comparison with the structure in air, the triangular-like hole array in the (111) plane was found rotated ~30° in the distorted SU-8 structure since the motifs of the two structures were different. It was suggested that the rotation was caused by the interference terms between the three oblique surrounding beams, which disappeared in the air since the three polarization vectors of the side beams were pairwise perpendicular to each other.[14]

Comparison between theoretical values and experimental results

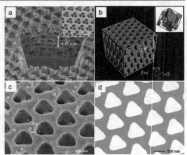

Figure 1. (a) Close-up SEM image of the fabricated SU-8 structure with the top surface tilted by 30°. The cross-section is FIB milled perpendicular to the (111) plane. The green dotted lines indicate four adjacent (111) lattice planes. Taking into account the viewing angle (60°), the distance between the adjacent lattice planes in the [111] direction is $h_{111}=0.81$ μm. The top (111) plane is partially melted by the ion beam. Inset: the (111) plane before FIB milling. (b) Distorted SU-8 structure described by Equation (3) with a filling fraction of 48% ($t'_{SU8}=0$) by considering both refraction and film shrinkage. The top surface is the (111) plane, the left one is the ($11\bar{2}$) plane, and the right one is the ($\bar{1}10$) plane. The unit length of the frame is $d = 1.38$ μm. Inset: the unit cell. (c) The bottom layer of the SU-8 structure. (d) 2D cut of the (111). Reprinted with permission from ref. 14. Copyright © 2007 Optical Society of America.

To validate the theoretical prediction and investigate the contributions of refraction and possible film shrinkage to the structure distortion, the theoretical values was compared with the experimental results [Figure 1]. The shrinkage is found to be ~ 1% in the (111) plane, and ~ 41% in the [111] direction. The results agree well with the report from a slightly different SU-8 structure patterned by HL.[21] The significant difference between experiment and theory may be attributed to the large shrinkage of SU-8 during the lithographic process, including photopolymerization and solvent development steps.[21] The shrinkage problem is well-known in the application of negative-tone resists (e.g. epoxy and acrylates).[22] Upon exposure to light, the low-molecular weight SU-8 resins polymerize and crosslink into an infinite network based on a cationic ring opening reaction, leading to shrinkage in the film. The percentage of shrinkage depends on the level of epoxy functionality involved in photocrosslinking, the exposure dosage and post-exposure bake time and temperature. In addition, during development, the organic solvent would swell the crosslinked film, which further distorts the patterned film after drying. Since the bottom layer of SU-8 film was confined by the substrate while the perpendicular direction was free, it is not surprising that the 3D structure shrank anisotropically.

Considering both refraction effect and resist shrinkage with respect to the resulting HL polymer structure, we have reconstructed a new level surface,[14]

$$F'_{SU8}(\vec{r}) = \cos[\frac{2\pi}{d}(-1.006x + 0.994y + 0.994z)] + \cos[\frac{2\pi}{d}(0.994x - 1.006y + 0.994z)]$$
$$+ \cos[\frac{2\pi}{d}(0.994x + 0.994y - 1.006z)] + 0.301\cos[\frac{2\pi}{d} \cdot 2(x-y) - \frac{\pi}{3}] \qquad (3)$$
$$+ 0.301\cos[\frac{2\pi}{d} \cdot 2(y-z) - \frac{\pi}{3}] + 0.301\cos[\frac{2\pi}{d} \cdot 2(z-x) - \frac{\pi}{3}]$$

where $d = 1.38$ μm. The new level surface matches well with experimental results, showing a noninversion symmetric structure that belongs to the space group No.155 ($R32$)[19,20] [Figure 1(b)]. The lattice parameters of the structure were calculated and presented in Table 2. We found that after considering both refraction effect and film shrinkage, the reconstructed SU-8 structure is close to the f.c.c lattice in air except that the triangular-like hole array in the (111) plane rotates by ~30° [Figure 1 (c) and (d)]. This implies that the resist shrinkage nearly compensates the lattice stretching due to refraction of the wave vectors but not polarization vectors.

Influence of structure distortion on photonic bandgap (PBG) properties

Using the MIT Photonic-Band Package (MPB)[23] we calculated the PBGs of inverse 3D Si (dielectric constant $\varepsilon_r \sim 13$) photonic crystals templated from the corresponding polymer structures to investigate the effect of structure distortion. As seen in figure 2, the Si PCs templated from the distorted SU-8 structure had a decreased maximum quality factor (QF = gap/mid-gap ratio), from 5.3% to 1.5% for the complete PBG between the 2nd and 3rd bands, and from 3.7% to less than 0.2% for another gap between the 7th and 8th bands.

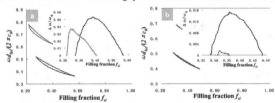

Figure 2. PBG maps of silicon PCs templated from (a) three-term diamond-like structure [Equation (1)] and (b) distorted structure by taking into account of both refraction and film shrinkage [Equation (3)]. Insets: quality factor vs. filling fraction. Black lines: the gap between the second and third band, and blue lines: the gap between the seventh and eighth band. $d_{fcc}=1.38$μm and c_0 is the light velocity in vacuum. Reprinted with permission from ref. 14. Copyright © 2007 Optical Society of America.

Design of optical setup to minimize distortion of SU-8 three-term diamond-like structures

To precompensate the unidirectional shrinkage, one has to design a spatially elongated interference intensity profile, which, after photoexposure and development, will result in a photonic crystal template with desired structural symmetry. For example, we know that the SU-8 film would shrink ~ 41% along the [111] direction. Assuming the crosslinking density in bulk and the shrinkage within the film is uniform, we can reconstruct the level surface for the elongated interference pattern as:

$$F_{\text{elongated}}(\vec{r}) = \cos[\frac{2\pi}{d}(-1.137x + 0.863y + 0.863z)] + \cos[\frac{2\pi}{d}(0.863x - 1.137y + 0.863z)]$$
$$+ \cos[\frac{2\pi}{d}(0.863x + 0.863y - 1.137z)] \tag{4}$$

where d is the lattice constant of the fcc lattice for three-term diamond-like structure.

Table 3. Wave vectors and polarization vectors in air, SU-8 using IL with visible light ($\lambda = 532$ nm).

Medium	Refraction Beams in SU-8	Incident Beams in Air
Refractive index	1.6	1.0
Wave vectors	$\vec{k}_0 = \frac{2\pi * 1.6}{\lambda}[\frac{1}{\sqrt{3}}, \frac{1}{\sqrt{3}}, \frac{1}{\sqrt{3}}];$	$\vec{k}_0 = \frac{2\pi}{\lambda}[\frac{1}{\sqrt{3}}, \frac{1}{\sqrt{3}}, \frac{1}{\sqrt{3}}];$
	$\vec{k}_1 = \frac{2\pi * 1.6}{\lambda}[0.856, 0.366, 0.366];$	$\vec{k}_1 = \frac{2\pi}{\lambda}[0.966, 0.183, 0.183];$
	$\vec{k}_2 = \frac{2\pi * 1.6}{\lambda}[0.371, 0.852, 0.371];$	$\vec{k}_2 = \frac{2\pi}{\lambda}[0.183, 0.966, 0.183];$
	$\vec{k}_3 = \frac{2\pi * 1.6}{\lambda}[0.371, 0.371, 0.852].$	$\vec{k}_3 = \frac{2\pi}{\lambda}[0.183, 0.183, 0.966].$
Polarization vectors for maximizing contrast for triple exposure	$\vec{E}_0 = E[0.408, 0.408, -0.817];$	$\vec{E}_0 = 1.300E[0.408, 0.408, -0.817];$
	$\vec{E}_1 = E[-0.065, -0.626, 0.777];$	$\vec{E}_1 = 1.453E[-0.031, -0.620, 0.784];$
	$\vec{E}_2 = E[-0.626, -0.065, 0.777];$	$\vec{E}_2 = 1.453E[-0.620, -0.031, 0.784];$
	$\vec{E}_3 = E[0.605, 0.605, -0.518].$	$\vec{E}_3 = 1.396E[0.683, 0.683, -0.258].$

The level surface described by equation 4 can be considered as a combination of three gratings, each of which is generated by the interference between the central beam and one side beams. It is critical that each set of beams should not interfere with the other during the exposure process, that is, the polarizations between any two side beams must be perpendicular to each other in a single exposure. Alternatively, the exposures can be separated in time by using triple exposures of two beams, one from the central beam and the other from the side beam each time.

In comparison with the single exposure of four-beam IL, multi-exposure of two-beam IL has more flexibility to generate various structures with scalable size since any arbitrary structure can be expanded as a sum of Fourier terms each of which could correspond to a grating arising from the interference of two beams.[13,24] Moreover, the two-beam multi-exposure can only keep the useful interference terms, and eliminate the unwanted ones. For instance

For the diamond-like structure, we can add three separate sets of high speed shutters on the optical paths of the side beams in the existing optical setup to perform triple exposures of the central beam and one side beam. The exposure time and interval between exposures can be controlled by shutters. All the wave vectors and polarization parameters are kept the same as those shown in Table 1. The total intensity profile can be expressed as the sum of all exposures:

$$I(\vec{r}) = \sum_{l=1}^{3} I_l = 3(E_0^2 + E_s^2) + 2 \cdot 0.659 E_0 E_s \{\cos[\frac{2\pi}{d}(-1.140x + 0.860y + 0.860z)]$$
$$+ \cos[\frac{2\pi}{d}(0.860x - 1.140y + 0.860z)] + \cos[\frac{2\pi}{d}(0.860x + 0.860y - 1.140z)]\} \tag{5}$$
$$= I_0 + \Delta I(\vec{r})$$

where I_l means the interference intensity of the lth exposure.

The only concern could be the increased background intensity and thus decreased the contrast of interference pattern in comparison with that from the four-beam single exposure, which is given by $C \equiv \max(\Delta I(\vec{r}))/I_0 \propto E_0 E_s / (E_0^2 + E_s^2)$ according to equation (4). The maximum ratio of $E_0 E_s/(E_0^2 + E_s^2) = 0.5$ could be achieved by setting $E_0 = E_s = E$. In a separate vein, the trial-and-error method suggests that another set of polarization vectors (in SU-8) [Table 3], which are all linear, can provide a high contrast intensity profile as:

$$I(\vec{r}) = 6E^2 + 2 \cdot 0.916E^2 \{ \cos[\frac{2\pi}{d} (-1.137x + 0.862y + 0.862z)]$$

$$+ \cos[\frac{2\pi}{d} (0.862x - 1.137y + 0.862z)] + \cos[\frac{2\pi}{d} (0.862x + 0.862y - 1.137z)]\} \qquad (6)$$

The optical system can be further simplified using a rotational sample stage [Scheme 1] to control the three exposures via rotating the sample around the c-axis by an angle of 120° between every two exposures.[25-28] The interference profile in SU-8 [Equation (4)] may be obtained by utilizing only two laser beams [Scheme 1] and the high contrast between the minimal and maximal intensities of interference pattern is accessible simply by setting the two laser beams with the same polarized direction and the same intensity.

Scheme 1. Illustration of Two-beam IL with the sample fixed on a rotation stage.

It is important to note that, if prisms and/or index matching liquids are not used to precompensate the effects of refraction, it is necessary to convert the calculated wave vectors and polarization vectors from SU-8 into the air by applying Snell's law and Fresnel's equations.[15] All parameters are summarized in Table 3. From the wave vectors in air, we can obtain the angle between the two laser beams $\theta = 39.8°$.

CONCLUSIONS

We have quantitatively studied the distortion of a three-term diamond-like structure fabricated by four-beam HL from SU-8 resist, which can be attributed to 1) refraction at the air-film interface, and 2) resist film shrinkage during lithographic process. To understand the effect of refraction, we compared the interference intensity profile and reconstructed level surfaces in SU-8 ($n \sim 1.6$) with those in air ($n=1$). Our calculation suggested that the SU-8 lattice was stretched along the [111] direction but no distortion in the (111) plane in comparison with that in air, and the symmetry of their level surface decreased from space group No.166 ($R\bar{3}m$) in air to No.155 ($R32$) in SU-8. Therefore, the translational symmetry was decreased from f.c.c in air to rhombohedral in SU-8. It also suggested that the SU-8 pattern would rotate by $\sim 30°$ in the (111) plane away from that in air due to the effects of refraction on polarization. The calculation was confirmed by experimental observation. We calculated the PBGs of the inversed 3D Si PCs templated from the distorted SU-8 structures, and observed decrease of the quality factors. Furthermore, to minimize the distortion, we have theoretically designed a new optical setup, two-beam triple exposure, to eliminate the interference of the side beams and generate an elongated interference profile, which after the unidirectional shrinkage, will result in the desired structural symmetry. The presented quantitative study in three-term diamond-like SU-8 films can be applied to many other 3D polymer templates patterned by HL.[29] It offers the first important step toward rationale design of appropriate optical systems for desired photonic crystals with large complete bandgaps.

ACKNOWLEDGMENTS

This work is supported by the Office of Naval Research (ONR), Grant # N00014-05-0303. XZ would like to thank Jun Hyuk Moon (Sogang Univ.), Jingjing Li (HP) and Liang Fu (Univ. of Pennsylvania) for useful discussion and insights.

REFERENCES

[1] E. Yablonovitch, Phys. Rev. Lett. **58**, 2059 (1987).

[2] S. John, Phys. Rev. Lett. **58**, 2486 (1987).

[3] J.D. Joannopoulos, S.G. Johnson, R.D. Meade, and J.N. Winn, *Photonic crystals*, 2nd ed. (Princeton University Press, 2008).

[4] S. Y. Lin, J. G. Fleming, D. L. Hetherington, B. K. Smith, R. Biswas, K. M. Ho, M. M. Sigalas, W. Zubrzycki, S. R. Kurtz, and J. Bur, Nature **394**, 251 (1998).

[5] S. Kawata, H. B. Sun, T. Tanaka, and K. Takada, Nature **412**, 697 (2001).

[6] M. Campbell, D. N. Sharp, M. T. Harrison, R. G. Denning, and A. J. Turberfield, Nature **404**, 53 (2000).

[7] Y. A. Vlasov, X. Z. Bo, J. C. Sturm, and D. J. Norris, Nature **414**, 289 (2001).

[8] S. R. Kennedy, M. J. Brett, O. Toader, and S. John, Nano Letters **2**, 59 (2002).

[9] K. K. Seet, V. Mizeikis, S. Matsuo, S. Juodkazis, and H. Misawa, Adv. Mater. **17**, 541 (2005).

[10] C. K. Ullal, M. Maldovan, M. Wohlgemuth, and E. L. Thomas, J. Opt. Soc. Am. A-Opt. Image Sci. Vis. **20**, 948 (2003).

[11] R. L. Sutherland, V. P. Tondiglia, L. V. Natarajan, S. Chandra, D. Tomlin, and T. J. Bunning, Opt. Express **10**, 1074 (2002).

[12] Y. V. Miklyaev, D. C. Meisel, A. Blanco, G. von Freymann, K. Busch, W. Koch, C. Enrich, M. Deubel, and M. Wegener, Appl. Phys. Lett. **82**, 1284 (2003).

[13] C. K. Ullal, M. Maldovan, E. L. Thomas, G. Chen, Y. J. Han, and S. Yang, Appl. Phys. Lett. **84**, 5434 (2004).

[14] X. Zhu, Y. Xu, and S. Yang, Opt. Express **15**, 16546 (2007).

[15] M. Born and E. Wolf, *Principles of optics : electromagnetic theory of propagation, interference and diffraction of light* (Cambridge University Press, Cambridge, U.K.; New York, 1999).

[16] Y. C. Zhong, S. A. Zhu, H. M. Su, H. Z. Wang, J. M. Chen, Z. H. Zeng, and Y. L. Chen, Appl. Phys. Lett. **87**, 061103 (2005).

[17] J. H. Moon, S. Yang, W. T. Dong, J. W. Perry, A. Adibi, and S. M. Yang, Opt. Express **14**, 6297 (2006).

[18] L. Z. Cai, X. L. Yang, and Y. R. Wang, Opt. Lett. **27**, 900 (2002).

[19] Th. Hahn, *International tables for crystallography. Vol A, Space-group symmetry*, 5th ed. ed. (Kluwer, Dordrecht, London, 2002).

[20] U. Shmueli, *International tables for crystallography. Vol. B, Reciprocal space*, 2nd ed. (Kluwer, Dordrecht, Boston, 2001).

[21] D. C. Meisel, M. Diem, M. Deubel, F. Perez-Willard, S. Linden, D. Gerthsen, K. Busch, and M. Wegener, Adv. Mater. **18**, 2964 (2006).

[22] W. H. Zhou, S. M. Kuebler, K. L. Braun, T. Y. Yu, J. K. Cammack, C. K. Ober, J. W. Perry, and S. R. Marder, Science **296**, 1106 (2002).

[23] S. G. Johnson and J. D. Joannopoulos, Opt. Express **8**, 173 (2001).

[24] J. H. Jang, C. K. Ullal, M. Maldovan, T. Gorishnyy, S. Kooi, C. Y. Koh, and E. L. Thomas, Adv. Funct. Mater. **17**, 3027 (2007).

[25] R. C. Gauthier and K. W. Mnaymneh, Opt. Laser Technol. **36**, 625 (2004).

[26] N. D. Lai, W. P. Liang, J. H. Lin, C. C. Hsu, and C. H. Lin, Opt. Express **13**, 9605 (2005).

[27] Y. Liu, S. Liu, and X. S. Zhang, Appl. Opt. **45**, 480 (2006).

[28] A. Dwivedi, J. Xavier, J. Joseph, and K. Singh, Appl. Opt. **47**, 1973 (2008).

[29] A. Hayek, Y Xu, T. Okada, S. Barlow, X Zhu, J. H. Moon, S. R. Marder, and S. Yang, J. Mater. Chem. **18**, 3316 (2008).

Mater. Res. Soc. Symp. Proc. Vol. 1182 © 2009 Materials Research Society   1182-EE13-14

Optical Feedback on Erbium Doped Fibre Laser for
Heterodyne SNOM Imaging Near 1.5μm

Pierre Bahin[1], Hervé Gilles[1], Sylvain Girard[1], Fabrice Gourbilleau[2], Mathieu Laroche[1] and Richard Rizk[2]
[1]LIOA Lasers, Instrumentation Optique et Applications
[2] NIMPH Nanostructures Intégrées pour la Microélectronique et la Photonique
Centre de recherche sur les Ions, les Matériaux et la Photonique (CIMAP), UMR 6252 CEA-CNRS-ENSICAEN, Université de Caen, 6 Boulevard Maréchal Juin, 14050 Caen Cedex 4
France

ABSTRACT

Heterodyne optical feedback on a class B laser is investigated for Scanning Near field Optical Microscopy (SNOM). All-fiberized set-up combining an Er-doped Distributed Feedback (DFB) fiber laser, a pair of pigtailed acousto-optics modulators (AOM) and a shear-force based scanning probe technique has been developed for the simultaneous observation of topography and evanescent light field on integrated optical devices. First demonstration of imaging using this technique is illustrated by characterizing the propagating modes into a rib waveguide at 1.54μm. Comparison between a theoretical model based on beam propagation mode (BPM) simulations and experimental measurements validates the results.

INTRODUCTION

Since the development of Scanning Probe Microscopy at the early 80's, a lot of technologies have been successfully implemented depending on the physical interactions between the scanning microprobe and the surface of the studied object and/or the geometry of the device. For characterizing integrated optical devices (waveguide, micro-structured fibres, surface plasmon resonance, photonics band gap materials...), Scanning Near Field Optical Microscopy (SNOM) combined with Atomic Force Microscopy (AFM) were early recognized as the most efficient and reliable characterization techniques. Combining AFM/SNOM measurements permit to characterize simultaneously the topography and the optical functionality of sub-micrometric integrated optical components. However, as the optical power collected or diffracted by the micro tip remains extremely small (typically a few tens of pW or even less), most of the classical SNOM devices works in the visible domains with high signal/noise ratio detectors like silicon photodiodes or photomultiplier tubes. It remains relatively difficult to extend the technology above 1μm as the performance of the photoelectric detectors decreases significantly in the near infrared spectral domain.

Recently, significant improvements on SNOM technology have been obtained thanks to the heterodyne coherent detection approach [1,2]. This latter coupled to the SNOM (also called h-SNOM) consists in recovering simultaneously the amplitude and phase of the local optical field collected by the scanning microprobe for a full reconstruction of the evanescent wave properties. In h-SNOM approach, a Mach-Zehnder interferometer is usually used to mix the optical field collected by the microprobe with a frequency shifted reference optical beam. A lock-in amplifier, referenced on the frequency shift, allows recovering simultaneously the amplitude and phase on the photo-detected signal. Moreover, the heterodyning permits a significant improvement considering the sensitivity and dynamical range of the measurement system as the collected electric field E_d is multiplied by the reference field E_{ref} in the detected signal. Several groups investigated this approach to develop h-SNOM instruments near 1.5µm [3-5] - a spectral domain extremely useful for characterizing integrated optical components with functionalities for the telecom market. However, it remains difficult to use extremely small micro tip (less than 20nm) in the infrared domain, as the signal-to-noise decreases proportionally to the optical power P_{tip} collected by the micro tip. Any further improvement of the sensitivity in the h-SNOM approach would be therefore of great interest.

An original way for coherent detection of extremely low photon flux consists in using the optical feedback on a class B laser source. The Laser Feedback Interferometry (LFI) technique has been already successful for coherent imaging inside turbid media [6]. The selection of a frequency shift near the relaxation oscillation of the laser creates an amplified beating note inside the laser, resulting in a very simple technique for obtaining quantum-noise limited detection without compromise on the performance of the optical detector [7]. In this case, the photo-detected signal Sd becomes:

$$Sd = E_O^2 + E_{probe}^2 + 2\gamma E_O E_{probe} \times \cos(2\pi\Delta f \cdot t + \Phi)$$

Where E_0 is the output electric field of the laser, E_{probe} is the evanescent field collected by the micro tip, Δf is the frequency shift and γ is an enhancing factor which could reach 10^6 for Δf close to the relaxation oscillation frequency of the laser [7].

In a previous paper [8], coherent heterodyne feedback on class B solid-state laser was already suggested for the detection of evanescent optical field. The present paper extends the technique and demonstrates that LFI-SNOM can be effectively used for near-field imaging on integrated optical devices, following the suggestion mentioned in reference [8] and recently investigated by another group [9]. 2D-near field optical imaging was done on a guided-mode in a Si-rich rib waveguide and the results are numerically analyzed using BPM software.

EXPERIMENTAL SET-UP AND RIB WAVEGUIDE CHARACTERISTICS

The experimental set-up used for 2D-imaging of the evanescent optical field based on heterodyne optical feedback on a class-B laser is illustrated on figure 1.

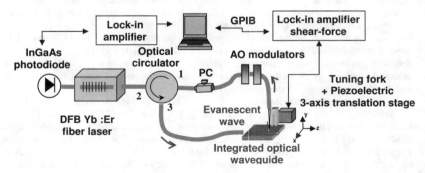

Figure 1: Experimental set-up of Laser Feedback Interferometer for Scanning Near Field Microscope.

This set-up is a direct evolution of the initial device (see reference [8]). The stability has been further improved by replacing most of the bulk optical components by fiberized equivalents. The laser source is now an erbium doped DFB fibre laser (Koheras Basik C15) emitting up to 8mW in single-longitudinal mode at λ=1535nm, directly spliced to an optical circulator used as a beam splitter/recombiner. The laser relaxation oscillation frequency is typically near 700 kHz. In order to couple the light into the studied rib waveguide, the injection fiber is directly butted on the entrance facet of the waveguide.

The guiding layer is made of 1.2µm thick silica layer containing Er^{3+} ions and Si nanoclusters, obtained after growth by reactive magnetron sputtering and then annealing at 900°C for 1h under nitrogen flux. It was deposited on a bottom cladding SiO_2 layer (5µm thick) previously thermally grown on Silicon wafer. Initial designs using BPM software suggest that 5µm thick bottom cladding should be enough to avoid leakage losses into the Si substrate. This point will be further discussed in the next section of the paper. Optical lithography and reactive ion etching have been further used to process a series of 2-3 cm long rib-loaded waveguides with rib depth of 350nm and widths ranging between 1.5 to 10µm. More details on the fabrication processes can be found in [10, 11]. No top cladding was added on the current sample, allowing easier detection of the evanescent wave via SNOM photon-tunneling detection. A scheme of the waveguide structure is shown on figure 2.a.

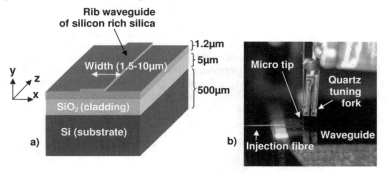

Figure 2: a) Scheme of the studied ridge waveguides. The width of the ribbon varies between 1.5µm and 10µm but most of the scans were done on 7µm wide ribbons; b) Image of the micro tip glued on the quartz tuning fork.

The optical fiber micro tip is approached along the y-axis (see figure 2.b) close to the air-waveguide interface – typically at a few tens of nm - via a 3-axis piezoelectric translation stage (Melles Griot 3D-Nanomax 17MAX301). The micro tip converts the evanescent wave into a collected propagating field. The tip is glued on a quartz tuning fork for detecting the evanescent wave at constant distance between the extremity of the micro-tip and the sample. The implemented home-made device is based on the well-known shear-force approach [12]. During a 2D transverse scan along the interface, the quartz tuning fork piezoelectric response can be easily used as an error signal for a closed-loop controller to maintain a constant distance between the micro-tip and the surface. The light collected by the micro tip is frequency shifted on the feedback branch by a pair of fibre-pigtailed acousto-optics modulator (AOMs) before being re-injected into the laser via the port 1 of the circulator. Finally, in order to detect the resulting dynamical perturbation, the rear output of the laser is injected through an optical isolator to avoid any spurious perturbation due to back-reflection. And the signal is detected using an InGaAs photodiode.

All the measurement instruments (the SNOM and shear-force lock-in amplifiers, the piezoelectric translation stage driver and the synthesizer for excitation of the quartz tuning fork) are computer-controlled via GPIB interface. Home-made software controls the automated initial approach of the micro-tip as well as maintains the optical probe at constant distance over the 2D scan and records the information provided by the lock-in detections. Moreover, the voltage applied to the piezoelectric translation stage to regulate the waveguide-probe distance can also be used to reconstruct the topography of the object. A series of tests on a Si calibration grating (TGZ3 from NT-MDT) has validated the efficiency of the shear-force based distance regulation mechanism.

STUDY OF THE PROPAGATION MODE IN RIB WAVEGUIDE USING SHEAR-FORCE AND LFI-SNOM IMAGING TECHNIQUES

Examples of the simultaneous 2D images of the topography and mode distribution along a 7µm width waveguide are shown on figure 3. A comparison between the shear-force topography (figure 3.a) and the SNOM profile (figure 3.b) shows that the mode is laterally well-confined inside the rib waveguide with most of the injected light propagating into the fundamental mode along the waveguide. Moreover, the ribbon edges are clearly identifiable on the SNOM image. The amplitude fluctuations along the waveguide axis could be attributed to surface imperfections or interference pattern due to spurious standing wave effect due to back-reflection on the output facet of the waveguide (even keeping into account that the waveguide propagation losses upper to 3dB/cm should limit such perturbation effect).

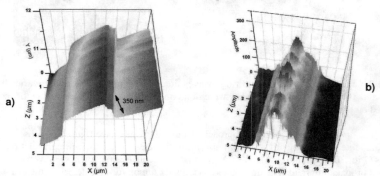

Figure 3: (a) Topography of a 7µm wide ridge waveguide obtained via the shear-force signal; (b) Electric field amplitude of the guided mode recorded on the top of the ribbon structure.

The second experimental observation done on the rib waveguides consists in measuring the mode cross section at the end of the waveguide directly in the near-field domain (Figure 4). Compared to the classical approach based on far-field imaging with a CCD beam profiler, the SNOM technique ensures a better dynamics (as the recorded signal is directly proportional to the electric field amplitude and not the intensity) and an adjustable transverse resolution depending on the micro-tip transverse size and/or scanning step. Moreover, it also gives access to potential losses due to evanescent leaks on the output facet of the waveguide.

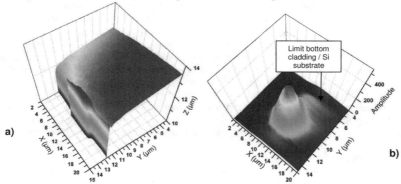

Figure 4: (a) Topography of the output facet of a 7μm wide ridge waveguide; (b) Electric field amplitude of the guided mode recorded at the output facet of the rib waveguide

On figure 4.b, it is easy to identify the position of the ribbon shape on one side of the mode profile, similar to the one observed on the shear-force profile. On the opposite side, it is also easy to distinguish the interface between the bottom cladding and the substrate (the straight line at the end of the arrow, 6μm below the ribbon). This observation suggests that the guided mode could present significant leakage losses into the Si substrate due to insufficient confinement.

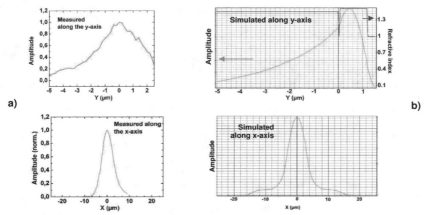

Figure 5: (a) Measured electric field profile of the propagation mode at the output facet of a 7μm wide rib waveguide; (b) Simulation obtained via BPM software.

In order to explain this last observation, additional BPM simulations were calculated (see figure 5.b) assuming a guiding layer refractive index equal to 1.507 (rather than 1.52 used before) and a bottom cladding refractive index equal to 1.45. This smaller refractive step index allows a good agreement between the numerical BPM results and the measured profile using the LFI-SNOM technique (figure 5). It suggests that a precise control of the refractive index value via the doping of silica with Si nanoclusters could be a relatively difficult task.

Finally, the results also reveal that the initial design of the waveguide – assuming higher refractive index contrast between the cladding and the guiding layer– was not optimized and causes leakage losses into the substrate. It could partially explain the high average distributed losses (up to 3dB/cm) observed during the optical characterizations of the waveguides series. Moreover, the long edge inside the cladding down to the substrate observed along the y-axis profile could be easily recorded using the LFI-SNOM technique because the detected signal is intrinsically proportional to the electric field. This last observation would be more difficult with an imaging approach based on CCD beam profiler, as it usually presents a limited dynamic.

CONCLUSIONS

A coherent detection based on frequency-shifted laser feedback interferometry (LFI) on Er-doped DFB fibre laser has been investigated for scanning near-field optical microscopy (SNOM). It is a powerful tool for characterizing and measuring optical properties on integrated photonic devices near 1.53µm with quantum-noise limited signal-to-noise ratio. As an example, the evanescent field topography recorded on the top (or at the end) of a rib waveguide has been tested using this technique.

Allowing the simultaneous determination of amplitude and phase of the evanescent field, the technique should now be also extended to observe amplitude-phase detection as it should exalt phase singularities due to propagation modes competition in waveguide structures.

REFERENCES

1. M.L.M. Balistreri, J.P. Korterik, L. Kuipers, N.F. van Hulst, *Physical Review Letters*, **85**, 294 (2000).
2. A. Nesci, R. Dändliker, H.P. Herzig, *Optics Letters*, **26**, 208 (2001).
3. A. Nesci, Y. Fainman, Proceedings of SPIE. **5181**, 62 (2003).
4. P. Tortora, M. Abashin, I. Märki, W. Nakagawa, L. Vaccaro. M. Salt, H.P. Herzig, U. Levy, Y. Fainman, *Optics Letters*, **30**, 2885 (2005).
5. I. Stefanon, S. Blaize, A. Bruyant, S. Aubert, G. Lerondel, R. Bachelot and P. Royer, *Optics Express*, **13**, 5553 (2005)
6. E. Lacot, R. Day, F. Stoeckel, *Optics Letters*, **24**, 744 (1999).
7. E. Lacot, R. Day, F. Stoeckel, *Physical Review* A, **64**, 438151 (2001).
8. H. Gilles, S. Girard, M. Laroche and A. Belarouci, *Optics Letters*, **33** , 1 (2008).
9. S. Blaize, B. Bérenguier, I. Stéfanon, A. Bruyant, G. Lérondel, P. Royer, O. Hugon, O. Jacquin and E. Lacot, *Optics Express*, **16** , 11718 (2008).
10. D. Navarro-Urrios, N. Daldosso, C. Garcia, P. Pellegrino, B. Garrido, F. Gourbilleau, R. Rizk and L. Pavesi, *Japanese Journal of Applied Physics*, **46**, 6626 (2007).
11 F. Gourbilleau, M. Levalois, C. Dufour, R. Rizk, *Journal of Applied Physics*, **95**, 3717 (2004).
12. K. Karrai, R.D. Grober, *Appl. Phys. Lett.* **66**, 1842 (1995).

Mater. Res. Soc. Symp. Proc. Vol. 1182 © 2009 Materials Research Society 1182-EE13-38

Subwavelength Terahertz Waveguide Using Negative Permeability Metamaterial

Atsushi Ishikawa,[1,2] Shuang Zhang,[1] Dentcho A. Genov,[1] Guy Bartal,[1] and Xiang Zhang[1,3]
[1]NSF Nanoscale Science and Engineering Center (NSEC), 5130 Etcheverry Hall, University of California, Berkeley, California 94720-1740, USA
[2]Japan Society for the Promotion of Science (JSPS) postdoctoral fellow for research abroad
[3]Materials Sciences Division, Lawrence Berkeley National Laboratory, 1 Cyclotron Road, Berkeley, California 94720, USA

ABSTRACT

We propose a novel subwavelength terahertz (THz) waveguide using the magnetic plasmon polariton (MPP) mode guided by a narrow gap in a negative permeability metamaterial. Deep subwavelength wave-guiding ($< \lambda/10$) with a modest propagation loss (2.5 dB/λ) and group velocities down to $c/21.8$ is demonstrated in a straight waveguide, a 90-degree bend, and a splitter. The distinctive dispersions of the guided mode with positive and negative group velocities are explained analytically by considering the dispersive effective optical constants of the metamaterial. The proposed waveguiding system inherently has no cutoff for any core width and height, paving the way toward the deep subwavelength transport of THz waves for integrated THz device applications.

INTRODUCTION

Due to the diffraction limit of light, the size of an optical beam cannot go beyond half of the wavelength in the free space, resulting in a significant challenge for miniaturization of optical-integrated devices and improvement of the spatial resolution in optical imaging. The main approach to overcome this limitation is the utilization of surface plasmon polaritons (SPPs), the surface modes present at metal-dielectric interfaces [1,2]. By exploiting the highly localized electromagnetic (EM) fields of SPP, a variety of plasmonic waveguides that efficiently transfer EM energy below the diffraction limit have been demonstrated in the optical frequency region [3-5].

Although the plasmonic structures are good candidates for the subwavelength devices in the optical frequency region, these promising systems, however, cannot be used at much lower frequencies. In the THz region where metal resembles a perfect electric conductor (PEC), the EM fields of SPP hardly penetrate into the metal, instead extend hundreds of wavelengths into the dielectric medium [6]. Although it has been demonstrated that SPP-like surface modes can be excited even on a PEC slab by introducing subwavelength holes on its surface [7,8], strong confinement and guiding of THz waves in the deep subwavelength scale ($< \lambda/10$) remains a major challenge and alternative approaches are highly desired.

The discovery of a new class of artificial materials, referred to as "metamaterials", has introduced the new means to engineer on demand the intrinsic properties of materials, i.e., the permittivity (ε) and the permeability (μ) [9,10]. These capabilities, supported by the advances in micro/nano fabrication technologies, bring various unprecedented optical functionalities into reality [11-14]. Of special interest are metamaterials with negative magnetic responses, which may offer us opportunities to manipulate the light more efficiently. [15-17]. In this paper, we propose a subwavelength THz waveguide using a negative permeability metamaterial as cladding between two parallel metal plates. We demonstrate that gap MPP modes guided by a vacuum core lead to the realization of two-dimensional (2D) strong confinement and guiding of THz waves in the deep subwavelength scale [18].

SUBWAVELENGTH THZ WAVEGUIDE USING NEGATIVE PERMEABILITY METAMATERIAL

Figure 1(a) shows the proposed structure of a subwavelength THz waveguide in the form of a rectangular vacuum core with a cladding made of a negative permeability metamaterial. The cladding is sandwiched between two parallel metal plates, providing the electric boundary condition to confine THz waves in the (x) vertical direction. In analogy to SPPs, negative permeability materials can support surface MPPs, which are transverse electric (TE) waves with the electric field E_x parallel to the material-vacuum interface [19]. When the gap width is small enough, the MPPs excited at each interface are coupled, resulting in symmetric and antisymmetric gap MPP modes, defined by the E_x symmetry along the y direction. The symmetric mode, which has no cutoff in the zero gap-width limit, can be used for the confinement in the y direction (perpendicular to E_x). The proposed structure inherently has no cutoff for any core width and height, thus realizing 2D strong confinement in the deep subwavelength core.

As shown in Fig. 1(b), we designed an array of metal-dielectric-metal magnetic resonators to realize the metamaterial that exhibits isotropic negative permeability in the y and z directions [16]. Each magnetic resonator consists of a high-dielectric material (ε_1) sandwiched between two metal plates, and the resonators are arranged in a three-dimensional array. Another low-dielectric material (ε_2) is used to separate the resonators in the x direction. The metamaterial was numerically investigated using a finite-difference time-domain (FDTD) software package (CST Microwave Studio). In the calculation, silver is chosen for metal (ε_m), and Ba$_2$Nd$_5$Ti$_9$O$_{27}$ and Benzocyclobutene, which are high- and low-permittivity dielectrics, are used as ε_1 and ε_2 media, respectively [20-22]. Figure 1(c) shows the real and imaginary parts of the retrieved

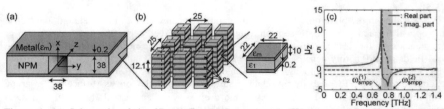

Figure 1. (a) Schematic of the 2D confined subwavelength THz waveguide. (b) An array of magnetic resonators is used to form a metamaterial that exhibits isotropic negative permeability in the y and z directions, and a single metal-dielectric-metal magnetic resonator with the dimensions indicated in μm. (c) Real (solid curve) and imaginary (dashed curve) parts of the retrieved μ_z of the metamaterial. The shaded region from $\omega_{smpp}^{(1)}$ to $\omega_{smpp}^{(2)}$ indicates the frequency range where $\mu_z < -1.0$.

effective permeability μ_z of the metamaterial. A magnetic resonance is clearly observed at 0.71 THz. In the shaded region from 0.71 to 0.85 THz, which are surface MPP resonant frequencies $\omega_{smpp}^{(1)}$ and $\omega_{smpp}^{(2)}$, the resonator array exhibits $\mu_z < -1.0$.

SUBWAVELENGTH CONFINEMENT AND GUIDING OF THZ WAVES

Once the magnetic response of the metamaterial has been characterized, we now study the frequency-dependent wave propagations in the proposed structure shown in Fig. 1(a). The core width and height (38.0 μm) are deep subwavelength compared to the operating wavelength (~ 400 μm). Figure 2(a) shows the numerically retrieved dispersion relations for both the metamaterial cladding without the core (circle dots) and the gap MPP mode of the waveguide (triangle dots). The normalized spectrum of transmitted electric field E_x through the waveguide is also shown in Fig. 2(b). A band gap opens in the dispersion of the metamaterial due to the negative magnetic response in the frequency range of 0.71 to 1.06 THz. Inside the band gap, three branches labeled as I, II, and III are formed by the gap MPP mode with dispersions closely following the analytical result (dashed curve). Three branches are divided by two surface MPP resonances at $\omega_{smpp}^{(1)}$ and $\omega_{smpp}^{(2)}$. In the shaded region where $\mu_z < -1.0$, the second branch, which contributes to the transmission peak centered at 0.77 THz, departs significantly from the light line and reaches asymptotically the upper edge of shaded region at large k_z. Note that at 0.85 THz, a large $k_z = 4.6\ k_0$ ($k_0 = \omega/c$ is the wavenumber in vacuum) is obtained, while a low group velocity of $c/21.8$ is achieved at 0.77 THz. Above the shaded region where $-1.0 < \mu_z < 0$, the third branch, which is a quasi-bound mode with a negative group velocity ($\partial\omega/\partial k_z < 0$) [23], folds back toward the upper edge of the band gap at small k_z. Similar dispersion property can be also found for the first branch just below the shaded region.

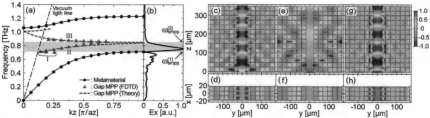

Figure 2. (a) Numerically retrieved dispersion relations for both the metamaterial cladding (circle dots) and the gap MPP mode of the waveguide (triangle dots). The analytical result for the gap MPP mode, dashed curve, is in good agreement with the FDTD result. No valid FDTD data were obtained above 0.9 THz because of weak transmission, while the analytical result predicts the existence of a fourth branch. The vacuum light line, dot-dashed line, is also shown for reference. (b) Corresponding transmission spectrum of the electric field E_x through the waveguide. The shaded region from $\omega_{smpp}^{(1)}$ to $\omega_{smpp}^{(2)}$ indicates the frequency range where $\mu_z < -1.0$. Magnetic field H_y distributions in the y-z plane at x = 0 μm (c) at 0.77 and (e) at 0.66 THz and those in the x-y plane at z = 260 μm (d) at 0.77 and (f) at 0.66 THz. Corresponding electric field E_x [(g) and (h)] in both the y-z and x-y planes at 0.77 THz. The metamaterial cladding in (c) - (h) indicated by the black lines consists of 3 x 6 x 14 unit cells in the x, y, and z directions.

To better understand the wave propagation inside the waveguide, we study the EM field distributions of the second branch at different frequencies. Figures 2(c) and 2(e) show the H_y field profiles in the y-z plane at x = 0 μm at 0.77 and 0.66 THz, corresponding to negative and positive permeabilities, respectively. At 0.77 THz, the symmetric H_y profile of the coupled MPPs excited at each interface is clearly observed and strong confinement in the deep subwavelength core is achieved. On the other hand, at 0.66 THz where the permeability is positive, the incident light diffuses into the claddings since the metamaterial no longer supports the MPPs. This subwavelength wave-guiding property is also seen in H_y distributions in the cross section of the waveguide at z = 260 μm for both frequencies [Figs. 2(d) and 2(f)]. At the immediate vicinity of the interface in Fig. 2(d), strong magnetic field associated with the gap MP mode is clearly observed and the coupled magnetic field is uniformly confined over the cross section of the core, which is in stark contrast to Fig. 2(f). In Figures 2(g) and 2(h), we also show the corresponding electric field E_x in both the y-z and x-y planes at the operating frequency of 0.77 THz. Similar confinement and wave-guiding properties are also obtained for the electric field. Following the confinement of the electric and magnetic fields, the EM energy flow in the waveguide is strictly concentrated in the core.

Despite the substantial imaginary part of μ_z at the considered frequency [Fig. 1(c)], a sufficient performance is still obtained since the absorption occurs only in a small area near the surface of the metamaterial claddings where the electric field is concentrated. For the considered configuration and operating frequency, we determined the propagation loss at 0.0064 dB/μm (2.5 dB/λ), which is on par with the subwavelength MPP transmission lines considered in the mid-infrared region [24]. Furthermore, the numerical studies indicate that the propagation loss mainly results from the absorption in the high-dielectric material and not in the metal, which can be improved further by using low-loss dielectrics or by optimizing the operating frequency.

SUBWAVELENGTH TRANSPORT OF THZ WAVES THROUGH SHARP BENDS

Since the core is surrounded by a negative permeability metamaterial and a PEC, there is no radiation mode to which the guided MPPs can couple; therefore, we expect that the subwavelength wave-guiding can be realized even in the presence of a sharp bend in the waveguide. Figures 3(a) and 3(b) show E_x distributions in the y-z plane of waveguides with a 90-degree bend and a splitter. In the calculations, the claddings are represented by an isotropic effective medium with μ = -7.79 + 0.86i at 0.77 THz. The core width is 50.0 □m and the other geometric parameters are the same as above. The subwavelength radiation is guided around the bend and splitter experiencing a low reflection. To quantitatively estimate the transmissivity through the bend, we consider two isometric waveguides. A comparison between the transmissions, with and without the bend, shows that the transmissivity through the bend reaches 96.5%. This is due to the improved impedance matching when the waveguide is sufficiently small compared to the wavelength and the bend can be considered as ajunction between two transmission lines with the same characteristic impedance [25]. In Figs. 3(c) and 3(d), we also show E_x distributions for the bend and splitter configurations by replacing the effective medium with the magnetic resonator array in the claddings. Although the wave is still guided around the bend, scattering is observed at the top corner where the electric field is strongly concentrated. As a result, only 23.8% transmissivity through the bend is observed. In the case of the splitter, the wave is well guided by the flat face of the resonator array and no significant scattering is observed at the T-junction. Surprisingly, almost 63.5% impinging energy is transferred along the wings of the sharp T-junction. Since the splitter, which can be considered as a junction of three transmission lines with the same characteristic impedance Z, has an intrinsic reflection loss $R = \left|(2Z - Z)/(2Z + Z)\right|^2 = 11.1\%$, the scattering loss in the system can be estimated at up to 25.4%. This result suggests that the scattering loss in the 90-degree bend geometry, which results from the imperfection of the resonator array at the sharp corner of the bend, can be improved by properly arranging the resonator array around the corner.

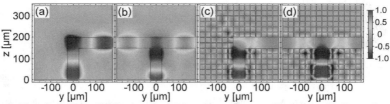

Figure 3. Electric field E_x distributions in the y-z plane of waveguides with a 90-degree bend [(a) and (c)] and a splitter [(b) and (d)] at 0.77 THz. The claddings in (a) and (b) are modeled as an isotropic effective medium with $\mu = -7.79 + 0.86i$, whereas those in (c) and (d) consist of the magnetic resonator array, indicated by the black lines.

CONCLUSIONS

We have proposed a subwavelength THz waveguide based on the gap MPP mode guided by a vacuum core in a negative permeability metamaterial. Strong confinement and guiding of THz waves below the diffraction limit are demonstrated in a straight waveguide, a 90-degree bend, and a splitter. These unique wave-confining features based on unconventional EM configurations could open new route to highly efficient subwavelength THz devices.

ACKNOWLEDGMENTS

This work was supported by the U. S. Department of Energy (DOE) under Contract No. DE-AC02-05CH11231 and partly by the U. S. Army Research Office (ARO) MURI program 50432-PH-MUR. A. Ishikawa acknowledges support from the Japan Society for the Promotion of Science (JSPS).

REFERENCES

1. J. Takahara, S. Yamagishi, H. Taki, A. Morimoto, and T. Kobayashi, *Opt. Lett.* **22**, 475 (1997).
2. E. N. Economou, *Phys. Rev.* **182**, 539 (1969).
3. S. A. Maier, P. G. Kik, H. A. Atwater, S. Meltzer, E. Harel, B. E. Koel, and A. A. G. Requicha, *Nature Mater.* **2**, 229 (2003).
4. S. I. Bozhevolnyi, V. S. Volkov, E. Devaux, J.-Y. Laluet, and T. W. Ebbesen, *Nature (London)* **440**, 508 (2006).

5. R. Oulton, V. Sorger, D. A. Genov, D. F. P. Pile, and X. Zhang, *Nat. Photon.* **2**, 496 (2008).
6. K. Wang and D. M. Mittleman, *Nature (London)* **432**, 376
7. S. A. Maier, S. R. Andrews, L. Martin-Moreno, and F. J. Garcia-Vidal, *Phys. Rev. Lett.* **97**, 176805 (2006).
8. C. R. Williams, S. R. Andrews, S. A. Maier, A. I. Fernandez-Dominguez, L. Martin-Moreno, and F. J. Garcia-Vidal, *Nat. Photon.* **2**, 175 (2008).
9. A. Ishikawa, T. Tanaka, and S. Kawata, *Phys. Rev. Lett.* **95**, 237401 (2005).
10. S. Zhang, W. Fan, N. C. Panoiu, K. J. Malloy, R. M. Osgood, and S. R. J. Brueck, *Phys. Rev. Lett.* **95**, 137404 (2005).
11. J. Valentine, S. Zhang, T. Zentgraf, E. Ulin-Avila, D. A Genov, G. Bartal, and X. Zhang, *Nature (London)* **455**, 376 (2008).
12. N. Fang, H. Lee, C. Sun, and X. Zhang, *Science* **308**, 534 (2005).
13. W. Cai, U. K. Chettiar, A. V. Kildishev, and V. M. Shalaev, *Nat. Photon.* **1**, 224 (2007).
14. Z. Liu, H. Lee, Y. Xiong, C. Sun, and X. Zhang, *Science* **315**, 1686 (2007).
15. T. J. Yen, W. J. Padilla, N. Fang, D. C. Vier, D. R. Smith, J. B. Pendry, D. N. Basov, and X. Zhang, *Science* **303**, 1494 (2004).
16. G. Dolling, C. Enrich, M. Wegener, J. F. Zhou, C. M. Soukoulis, and S. Linden, *Opt. Lett.* **30**, 3198 (2005).
17. T. Tanaka, A. Ishikawa, and S. Kawata, *Phys. Rev. B* **73**, 125423 (2006).
18. A. Ishikawa, S. Zhang, D. A. Genov, G. Bartal, and X. Zhang, *Phys. Rev. Lett.* **102**, 043904 (2009).
19. J. N. Gollub, D. R. Smith, D. C. Vier, T. Perram, and J. J. Mock, *Phys. Rev. B* **71**, 195402 (2005).
20. P. B. Johnson and R.W. Christy, *Phys. Rev. B* **6**, 4370 (1972).
21. P. Kuzel and J. Petzelt, *Ferroelectrics* **239**, 79 (2000).
22. Dow Chemicals: http://www.dow.com/cyclotene/solution/highfreq.htm.
23. J. A. Dionne, L. A. Sweatlock, H. A. Atwater, and A. Polman, *Phys. Rev. B*, **73**, 035407 (2006).
24. H. Liu, D. A. Genov, D. M. Wu, Y. M. Liu, J. M. Steele, C. Sun, S. N. Zhu, and X. Zhang, *Phys. Rev. Lett.* **97**, 243902 (2006).
25. G. Veronis and S. Fan, *Appl. Phys. Lett.* **87**, 131102 (2005).

Mater. Res. Soc. Symp. Proc. Vol. 1182 © 2009 Materials Research Society   1182-EE13-41

Improving the Near-Field Transmission Efficiency of Nano-Optical Transducers by Tailoring the Near-Field Sample

Kursat Sendur[1] and William Challener[2]
[1]Sabanci University, Istanbul, 34956, Turkey
[2]Seagate Technology, Pittsburgh, PA 15222, USA

ABSTRACT

Despite research efforts to find a better nano-optical transducer for light localization and high transmission efficiency for existing and emerging plasmonic applications, there has not been much consideration on improving the near-field optical performance of the system by engineering the near-field sample. In this work, we demonstrate the impact of tailoring the near-field sample by studying an emerging plasmonic application, namely heat-assisted magnetic recording. Basic principles of Maxwell's and heat transfer equations are utilized to obtain a magnetic medium with superior optical and thermal performance compared to a conventional magnetic medium.

INTRODUCTION

Emerging plasmonic nano-applications, such as heat-assisted magnetic recording (HAMR), require intense optical spots beyond the diffraction limit. Recent advances in near-field optics achieved spatial resolution significantly better than the diffraction limit. The possible ways to achieve intense optical spots with small sizes include apertures on good metallic conductors [1], bow-tie antennas [2], ridge waveguides [3], and apertureless near-field microscopy [4]. However, the power density requirements of emerging plasmonic applications, such as HAMR, are quite challenging for the nano-optical systems listed above. Therefore, further enhancement in power transmission is necessary.

Despite research efforts to find a better transducer for light localization and high transmission efficiency, there is not much consideration on improving the near-field optical performance of the system by engineering the near-field sample. Some emerging plasmonic applications, such as heat assisted magnetic recording, allow tailoring the shape and composition of the near-field sample. As we demonstrate in this work, a near-field sample has a significant impact on the nano-optical system performance when the sample is designed based on the fundamental principles of optical energy transfer and heat flux propagation.

In this work, we demonstrate the impact of tailoring the near-field sample by studying an emerging plasmonic application, namely heat-assisted magnetic recording [5,6]. The near-field sample in a heat-assisted magnetic recording system is the recording medium. In HAMR, the main requirement is that the temperature of the magnetic medium should be increased to the Curie temperature. Typical Curie temperatures of magnetic materials are on the order of several hundred degrees higher than the ambient temperature. Basic principles of Maxwell's and heat transfer equations are utilized to obtain a magnetic medium with superior optical and thermal performance compared to a conventional magnetic medium. The fundamentals of the optical energy coupling at the optical transducer-magnetic medium interface are explained via Maxwell's equations. Various optical and thermal aspects of the impact on the near-field sample are discussed. The optical performance of the tailored recording medium is compared with a conventional magnetic medium using 3-D finite element method solutions of Maxwell's and the heat transfer equations when a nano-optical transducer is illuminated with a focused beam of light defined by Richards-Wolf vector field equations. Based on the results, a patterned magnetic medium for HAMR is suggested to increase the optical energy transmission from a near-field transducer to the medium.

NEAR-FIELD TRANSDUCER-SAMPLE INTERACTION

To understand the optical energy transfer mechanism at the interface between a near-field transducer and a sample, the interaction of various electromagnetic field components produced by a transducer with a sample must be considered. In Fig. 1 a schematic illustration of a ridge waveguide transducer, which is placed at the end of an optical lens system, is provided. The optical transducer is placed in the vicinity of a near-field sample, which is the recording magnetic medium for HAMR. The electromagnetic field distributions at this interface have similarities for different transducers due to Maxwell's boundary conditions for good metals. In a previous study, the x, y, and z components of the electric field were presented in the magnetic medium when it was in the vicinity of a ridge waveguide [7]. Figures 3-5 in that study suggest that the strength of the perpendicular (i.e. z) component of the electrical field is comparable to or even larger than the transverse (i.e. x and y) components. Since the transducer acts as an electric dipole, it would be expected to produce significantly stronger transverse component than the perpendicular component. Within the tranducer, strong transverse field components are present. Outside the transducer, especially in the space between the transducer and the recording medium, a stronger perpendicular component is present. The presence of a stronger perpendicular component can be best understood by the boundary conditions on good metals. Good metals force the electric field lines to be perpendicular to their surface. Strong perpendicular electric field components are even more prominent for apertureless optical transducers [4]. Sample electric field distributions around transducers can be found in the literature [7-8].

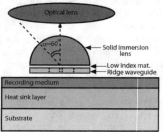

Figure 1. A schematic illustration of an optical transducer, which is placed at the end of an optical lens system, is given. The optical transducer is placed in the vicinity of a near-field sample, which is the recording magnetic medium for HAMR.

In light of this discussion, the main concern for practical plasmonic systems is to efficiently couple the strong perpendicular component of the electric field into the sample. Coupling of the electric fields into the recording medium determines the transmission efficiency of the system. The coupling of the electric fields at the interface between the transducer and the magnetic medium are determined by the boundary conditions. Maxwell's equations state that: (a) the tangential component of the electric field across an interface between two media is continuous, $\vec{E}_1^t = \vec{E}_2^t$, (b) the perpendicular component of the electric field intensity across an interface is discontinuous by an amount $\vec{E}_1^n / \vec{E}_2^n = \varepsilon_2 / \varepsilon_1$, where ε_1 and ε_2 are the permittivities of each material.

For different applications, the permittivity of the sample exhibits a large variation. In HAMR, the permittivity of magnetic materials at optical frequencies is significantly larger in magnitude than the permittivity of materials that fill the interface. The perpendicular component of the electric field in the magnetic medium is very low for the conventional magnetic medium. This results in low coupling efficiency for the perpendicular component of the electric field into a conventional magnetic medium. Since a strong perpendicular electric field is present at the interface between the transducer and magnetic medium, the discontinuity of the perpendicular component reduces the optical energy coupling from the transducer into the magnetic medium.

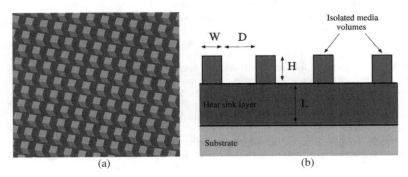

(a)   (b)

Figure 2. (a) An oblique view of the patterned magnetic medium with optically and thermally isolated magnetic particles placed over a heat sink layer, (b) Patterned magnetic medium with optically and thermally isolated magnetic particles placed over a heat sink layer. The geometric dimensions of the particle size and separation as well as the thicknesses of the magnetic and heat sink layers are shown.

RESULTS

The sample in the near-field systems can be designed so that the coupling of the perpendicular electric field component into the sample can be significantly improved. For example, in HAMR the magnetic medium, shown in Fig. 2, has a number of electromagnetic and thermal advantages over the conventional recording medium, including better electromagnetic field coupling and reduced thermal spread in the lateral direction. The designed medium, shown in Fig. 2, uses isolated magnetic volumes to increase both the light coupling and the thermal response. Isolated metallic volumes are made of magnetic material such as a CoPt alloy. Dielectric volumes are placed between these magnetic grains to isolate them both optically and thermally. The medium utilizes a highly conducting metallic heat-sink underlayer which quickly removes the heat from the medium. Fast heat removal is necessary for high recording data rates.

The improvement of localized heating in the magnetic medium shown in Fig. 2 is demonstrated using numerical simulations. Finite element method (FEM) 3-D electromagnetic and thermal modeling software is used. The finite element method (FEM) is a well-known numerical algorithm for the solution of Maxwell's equations [9]. In this study, a frequency-domain based FEM [10] is used for the solution of Maxwell's equations. Tetrahedral elements are used to discretize the computational domain, which allows modeling of arbitrarily shaped three-dimensional geometries. Over the tetrahedral elements, the edge basis functions and second-order interpolation functions are used to expand the functions. Adaptive mesh refinement is employed to improve the coarse solution regions with high field intensities and large field gradients. For thermal modeling, an FEM [10] based solution is used as well. FEM is a well-known and efficient computational technique, widely used for the solution of the heat transfer problems. In this study, we used commercially available FEM based thermal modeling software, which employs a time-domain-based FEM formulation for the solution of the heat transfer equation. FEM utilizes a Galerkin formulation, which minimizes the mean error over the computational volume. A combination of hexahedral, pentahedral, and tetrahedral elements are used to discretize the computational volume. Second-order interpolation functions are used to expand the unknown functions.

To illustrate the effect of the patterned medium shown in Fig. 2, we obtained 3-D finite element method solutions of Maxwell's and the heat transfer equations. The specifications of the numerical simulation are as follows. The transducer was illuminated with an optical power source of 100 mW. The recording medium model in this work is composed of uniformly-distributed same-size magnetic particles, which are equally separated from each other. The magnetic particles have side lengths of 5 nm. The separation between the particles is 5 nm and the height is 10 nm. The optical source produces about 30 nm spot size, which mainly illuminates the central 9 particles under the transducer. A heat sink layer with a 200 nm thickness is placed under the magnetic particles. The operating wavelength of the optical source is 700 nm. The thermal and optical properties of the magnetic layer and the heat sink layer are listed in Table 1. Figures 3 (a) and (b) illustrate the temperature distribution through a contour on the top surface of the recording medium. As shown in Figs. 3 (a) and (b), a higher temperature and a tighter localization are obtained by using a recording medium as shown in Fig. 2, instead of a conventional recording medium. Such high temperatures and small spots are essential for a practical HAMR system. Another potential area of interest for patterned near-field sample, which is described in this work, is optical data storage. Other examples of engineering near-field samples are discussed in the literature [11-12].

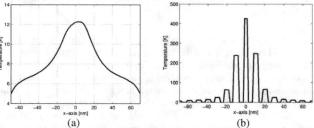

(a)   (b)

Figure 3. Temperature distribution in the (a) continuous recording medium and (b) patterned recording medium for a magnetic particle size of 5 nm and the duty cycle is 50%.

Table 1. Thermal and optical properties of the magnetic layer and the heat sink layer.

Layer	Density [W/m³]	Specific Heat [J/(kg K)]	Thermal Conductivity [W/(m K)]	Refractive Index [unitless]
Magnetic layer	8862	421	99.2	0.16 + i 3.95
Heat sink layer	19300	129	317	2.30 + i 4.43

The higher temperature and tighter localization shown in Fig. 3 (b) are primarily achieved by better optical coupling of electromagnetic waves into the patterned medium shown in Fig. 2 compared to conventional recording medium. In the case of a conventional recording medium, the electric field lines are perpendicular to the medium, as shown in Fig. 4 (a). However, in the case of patterned media, the electric field lines are perpendicular on the top surfaces and tangential on the side surfaces of the medium, as shown in Fig. 4 (b). The perpendicular component of the electric field intensity across an interface is discontinuous, but the tangential component of the electric field across an interface between two media is continuous. Due to this continuity, the tangential components of the field couple better to the medium. Better coupling of electromagnetic fields results in higher optical power absorption in the recording medium shown in Fig. 2.

In addition to better optical coupling, the patterned medium shown in Fig. 2 provides more favorable thermal conditions compared to a conventional recording medium. In the patterned medium shown in Fig. 2, the heat loss via lateral thermal conduction is greatly reduced because of the thermally insulating material between the magnetic particles. A decrease in lateral thermal conduction reduces the heat transfer to adjacent bits, and therefore, increases the temperature in the magnetic particle. The insulating material between the magnetic volumes also helps to prevent the thermal spread in the lateral direction, therefore, tighter thermal localization is achieved in Fig. 3 (b).

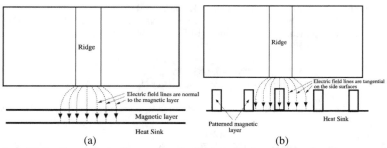

(a)   (b)

Figure 4. (a) Continuous recording medium is illustrated in the vicinity of a ridge waveguide. The electric field lines are normal to the continuous recording medium. (b) The electric field lines are both normal (at the top surfaces) and tangential (on the side surfaces) to the patterned media.

To achieve higher transmission efficiencies, and therefore, higher temperatures in the magnetic particles around the optical transducer, the sparsity S of the magnetic medium and the width of the magnetic particles should be reduced and the height H of the particles should be increased. This will make the media volumes more isolated and the electric field will better couple to the media due to the increase in the tangential component. Also the volumes become more thermally isolated, which will increase the temperatures. However, as previously mentioned, inappropriate selection of these parameters may result in a magnetically unstable medium. Therefore, these parameters should be optimized considering magnetic stability and the optical transducer performance.

The experimental conditions on the manufacturing of a patterned medium will also have an impact on the power transmission into the magnetic medium. The shape of the isolated media pattern volumes is an important factor determining the optical absorption in the recording medium. Particles with long and sharp side walls provide better optical absorption in the medium. Experimental imperfections, such as rough edges or undesired edge steepness, will reduce power transmission to the magnetic medium.

CONCLUSIONS

In summary, optical energy coupling at the interface between an optical transducer and near-field sample was explained via the boundary conditions of Maxwell's equations. Based on this description, the basic principles are applied to tailor the design of the near-field sample in HAMR. A patterned magnetic medium was suggested to increase the optical energy transmission from the near-field transducer to the medium. Results suggest that optically and thermally isolated magnetic particles can be used to increase the light coupling and temperature response in HAMR.

REFERENCES

1. F. J. G. de Abajo, Opt. Express 10, 1475 (2002).
2. R. D. Grober, R. J. Schoelkopf, and D. E. Prober, Appl. Phys. Lett. 70, 1354 (1997).
3. X. Shi and L. Hesselink, Jpn. J. Appl. Phys. 41, 1632 (2001).
4. A. Hartschuh, E. J. Sánchez, X. S. Xie, L. Novotny, Phys. Rev. Lett. 90, 095503 (2003).
5. T. McDaniel, W. Challener, and K. Sendur, IEEE Trans. Mag. 39, 1972 (2003).
6. R. Rottmayer et al., IEEE Trans. Mag. 42, 2417 (2006).
7. K. Sendur, W. Challener, and C. Peng, J. Appl. Phys. 96, 2743 (2004).
8. W. Challener et al., Jpn. J. Appl. Phys. 45, 6632-6642 (2006).
9. J. M. Jin, *The finite element method in electromagnetics* (John Wiley & Sons, New York, NY, 2000).
10. All the FEM calculations in this report are performed with software from Ansoft Inc. and Ansys Inc.
11. F. Hao et al., Phys. Rev. B 76, 245417 (2007).
12. D. W. Brandl et al., *Chem. Phys.* 123, 024701 (2005).

Mater. Res. Soc. Symp. Proc. Vol. 1182 © 2009 Materials Research Society   1182-EE13-42

An Integral Equation Based Numerical Solution for Nanoparticles Illuminated With Collimated and Focused Light

Kursat Sendur
Sabanci University, Istanbul, 34956, Turkey

ABSTRACT

An integral equation based numerical solution is developed when the particles are illuminated with collimated and focused incident beams. The solution procedure uses the method of weighted residuals, in which the integral equation is reduced to a matrix equation and then solved for the unknown electric field distribution. In the solution procedure, the effects of the surrounding medium and boundaries are taken into account using a Green's function formulation. Therefore, there is no additional error due to artificial boundary conditions unlike differential equation based techniques, such as finite difference time domain and finite element method. In this formulation, only the scattering nano-particle is discretized. The results are compared to the analytical Mie series solution for spherical particles, as well as to the finite element method for rectangular metallic particles. The Richards-Wolf vector field equations are combined with the integral equation based formulation to model the interaction of nanoparticles with linearly and radially polarized incident focused beams.

INTRODUCTION

Nano-optics is a rapidly growing field with a diverse set of existing and emerging practical applications [1-5]. A number of parameters have to be optimized in order to achieve large transmission efficiency while keeping the optical spot size well below the diffraction limit. Selecting an optimum set of parameters for a nano-optical transducer is important in achieving small spots and large transmission efficiencies. Due to the large number of geometry, material composition, and source-related parameters in nano-optical systems, the simulation times can be too large to optimize practical nano-optical systems.

To obtain accurate and fast computational solutions of nano-optical systems that involve a large number of geometry, material composition, and source-related parameters, the development of efficient and accurate modeling and simulation tools for near-field optical systems is necessary. In this study, an integral equation based numerical solution is developed for nano-optical particles when they are illuminated with collimated and focused incident beams. The numerical technique developed in this study requires only the discretization of the nano-optical transducer, rather than the entire structure. Therefore, it results in a fewer number of unknowns than the numerical algorithms currently being utilized for solutions of nano-optical systems, such as finite difference time domain and finite element method.

In this work, we provide a formulation for an integral equation based modeling and design tool for nano-optical systems. Similar tools have been successfully used for the analysis and design of other nano-optical systems in the literature. Nano-optical system modeling studies in the literature utilize differential equation based approaches, such as finite difference time domain (FDTD) [6-11] and finite element method (FEM) [11,12], as well as integral equation based techniques [13-20]. Previous integral equation based techniques have not presented three-dimensional results when the incidence excitation is composed of linearly and radially polarized tightly focused beams. A tightly focused beam of incident light provides a large incident electric field onto nanoparticles, improving the near-field radiation in the vicinity of the particle. Therefore, it is desirable to obtain integral equation based solutions when the incidence excitation is composed of linearly and radially polarized tightly focused beams. In this study, a three-dimensional integral equation based solution is obtained when the incidence excitation is composed of linearly and radially polarized tightly focused beams.

A full-wave implementation of the method of weighted residuals (MWR) [21-25], which is also known as the method of moments (MoM), has a number of advantages over FDTD and FEM for nano-optical system analysis. In MWR, the effects of the surrounding medium and boundaries are taken into account using a Green's function formulation. Therefore, MWR requires only the discretization of the nano-optical transducer, whereas FDTD and FEM require the discretization of the entire computational space. Therefore, the resulting matrix equations of the MWR are smaller in size. An additional advantage of an integral equation based approach is the reduction of the additional error due to the discretization of the boundaries. In an integral equation based approach, the boundary conditions are handled in Green's function formulation; therefore, there is no additional error due to the discretization of the boundaries. In a differential equation based approach, such as FDTD and FEM, however, there is additional error introduced into the solution due to artificial boundary conditions. In addition, the integration of complicated excitation functions, such as focused beams in a dense medium, is easier in an integral equation based MWR compared to FDTD.

In this work we provide a formulation of the integral equation based numerical solution. The integral equation is discretized into a matrix equation using the method of weighted residuals. In this study, the results of the numerical technique are compared to the results of the analytical Mie series solution for spherical particles and the finite element method for rectangular metallic particles. We also extended the formulation to the case where the incident excitation is defined as a focused beam of light. Richards-Wolf vector field equations are combined with the integral equation based formulation to model linearly and radially polarized focused beams.

METHOD OF WEIGHTED RESIDUALS

The total electric field is a result of the interaction of an incident optical beam with a nanoparticle. The total electric field $\vec{E}_{tot}(\vec{r})$ is composed of two components

$$\vec{E}_{tot}(\vec{r}) = \vec{E}_{inc}(\vec{r}) + \vec{E}_{scat}(\vec{r}) \tag{1}$$

where $\vec{E}_{inc}(\vec{r})$ and $\vec{E}_{scat}(\vec{r})$ are the incident and scattered electric field components, respectively. The incident electric field can be defined as the electric field propagating in space in the absence of a scattering object. The scattered electric field $\vec{E}_{scat}(\vec{r})$ in Eq. (1) represents the fields resulting from the interaction of the incident field $\vec{E}_{inc}(\vec{r})$ with the particles. In three-dimensional space, the scattered field $\vec{E}_{scat}(\vec{r})$ can be written as

$$\vec{E}_{scat}(\vec{r}) = \frac{i\omega\mu}{4\pi} \iint\limits_{S'} dS' \overline{\overline{G}}(\vec{r}, \vec{r}') \cdot \vec{J}(\vec{r}') \tag{2}$$

where $\vec{J}(\vec{r})$ is the induced current over the particle, ω is the angular frequency, μ is the permeability, and

$$\overline{\overline{G}}(\vec{r},\vec{r}') = \left[\overline{\overline{I}} + \frac{\nabla\nabla}{k^2}\right]\frac{e^{ik|\vec{r}-\vec{r}'|}}{|\vec{r}-\vec{r}'|} \tag{3}$$

is the dyadic Green's function in free space . To solve $\vec{J}(\vec{r})$, we will expand it into a summation

$$\vec{J}(\vec{r}) \cong \sum_{j=1}^{N} I_j \vec{b}_j(\vec{r}) \tag{4}$$

where $\vec{b}_j(\vec{r})$ represents known basis functions with unknown coefficients I_j. In this work, triangular rooftop basis functions are used to discretize the induced current over the nanoparticle. These basis functions are originally proposed by Glisson and Wilton [26] on rectangular domains and used on triangular domains by Rao et al. [27]. Triangular rooftop basis functions have been very popular due to their ability to model realistic geometries. Particle geometry is discretized in order to expand the induced current with triangular basis functions.

By utilizing the expansion given in Eq. (4), the electric field integral equation is obtained as

$$-\vec{E}_{inc}^{img}(\vec{r}) \cong \sum_{j=1}^{N} I_j \left\{ \hat{t} \cdot \frac{i\omega\mu}{4\pi} \iint_{S'} dS' \left[\overline{\overline{I}} + \frac{\nabla\nabla}{k^2}\right] \frac{e^{ik|\vec{r}-\vec{r}'|}}{|\vec{r}-\vec{r}'|} \cdot \vec{b}_j(\vec{r}') \right\} \tag{5}$$

Due to the approximation of the induced current with the summation in Eq. (4), there is a residual error in Eq. (5). The residual error in space can be written as

$$\Re(\vec{r}) = \vec{E}_{inc}^{img}(\vec{r}) + \sum_{j=1}^{N} I_j \left\{ \hat{t} \cdot \frac{i\omega\mu}{4\pi} \iint_{S'} dS' \left[\overline{\overline{I}} + \frac{\nabla\nabla}{k^2}\right] \frac{e^{ik|\vec{r}-\vec{r}'|}}{|\vec{r}-\vec{r}'|} \cdot \vec{b}_j(\vec{r}') \right\} \tag{6}$$

In the method of weighted residuals the error is distributed so that it is minimized in the minimum mean square sense. For this purpose, a new set of functions, known as weighting functions $\vec{w}_i(\vec{r})$ are used. The residual error $\Re(\vec{r})$ is distributed in space by equating the inner product of the residual error $\Re(\vec{r})$ with the weighting function $\vec{w}_i(\vec{r})$ to zero

$$\langle \Re(\vec{r}), \vec{\omega}_i(\vec{r}) \rangle = \int_{\Omega} \Re(\vec{r}) \cdot \vec{\omega}_i(\vec{r}) d\vec{r} = 0 \tag{7}$$

By placing the weighting functions into Eq. (7) we can obtain the resulting equations for the unknown coefficients of the basis functions. After mathematical manipulations, the result can be expressed as a system of linear equations as

$$\overline{\overline{Z}} \cdot \overline{I} = \overline{V} \tag{8}$$

where $Z_{i,j}$ is the impedance matrix element on the ith row and j^{th} column which is given as

$$Z_{i,j} = \frac{i\omega\mu}{4\pi} \left[\iint_{S} dS \vec{w}_i(\vec{r}) \cdot \iint_{S'} dS' \frac{e^{ik|\vec{r}-\vec{r}'|}}{|\vec{r}-\vec{r}'|} \vec{b}_j(\vec{r}') + \iint_{S} dS \vec{w}_i(\vec{r}) \cdot \frac{\nabla\nabla}{k^2} \cdot \iint_{S'} dS' \frac{e^{ik|\vec{r}-\vec{r}'|}}{|\vec{r}-\vec{r}'|} \vec{b}_j(\vec{r}') \right] \tag{9}$$

and V_i is the excitation source element on the ith row given as

$$V_i = -\iint_{S} dS \vec{w}_i(\vec{r}) \cdot \vec{E}_{inc}(\vec{r}) \tag{10}$$

By solving the matrix equation in Eq. (8), we obtain the unknown coefficients of the basis functions in the induced current expansion in Eq. (4).

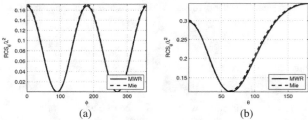

(a)   (b)

Figure 1. A comparison of the MWR results with the Mie series solution for the RCS of a conducting sphere with a radius of 140 nm. The operating wavelength is 700 nm. θ and ϕ components of the radar cross section are plotted on various cuts: (a) RCS_θ as a function of ϕ on $\theta=90°$ cut, (b) RCS_θ as a function of θ on $\phi=0°$ cut.

Using the integral equation based formulation given in the previous section, the interactions of a collimated beam with both a conducting metallic sphere and cube are studied. The collimated beam is modeled as a linearly polarized plane wave propagating in the z direction. In Fig. 1, the radar cross section of a sphere with a radius of 140 nm is presented to compare MWR results with the analytical Mie series solution. The operating wavelength of the laser source is 700 nm. A comparison of the MWR results with the analytical Mie series solution shows a good agreement between the results.

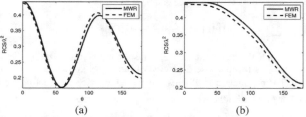

(a)   (b)

Figure 2. A comparison of the FEM and MWR results for the radar cross section of a conducting cube with a side length of 200 nm. The operating wavelength for the incident beam is 700 nm. θ component of the radar cross section is plotted on various ϕ cuts: (a) RCS_θ as a function of θ on $\phi=0°$ cut, (b) RCS_θ as a function of θ on $\phi=90°$ cut.

In Fig. 2, the scattering cross section of a conducting metallic cube with a side length of 200 nm is obtained on various cuts in the far-field. There is no analytical solution for a cube, therefore, we utilized an FEM solution as a reference. Similar to the previous calculations, a linearly polarized plane wave is utilized. The operating wavelength is 700 nm. In Fig. 2 (a) and (b) the θ component of the radar cross section is plotted on $\phi=0°$ and $\phi=90°$ cuts. The MWR and FEM results show a good agreement.

LINEARLY AND RADIALLY POLARIZED FOCUSED BEAMS

It is also very desirable to obtain solutions when the incidence excitation is composed of linearly and radially polarized focused beams. In the previous section, the integral equation based solutions are provided when the incident beam is a plane wave. In this section, the solution is obtained for the case where the incident beam is a focused beam. Richards and Wolf developed a method for calculating the electric field semi-analytically near the focus of an aplanatic lens [28, 29]. Using the Richards-Wolf method, we can obtain both transverse and longitudinal components near the focus for both linear and radial polarizations.

In Fig. 3 (a), various components of the near-field radiation from a sphere are plotted when the incident beam is a linearly polarized focused beam obtained from an optical lens system with a numerical aperture of 0.85. The operating frequency is 700 nm. The results are plotted for spherical particles with radii 70 and 140 nm. The E_x and E_z components are plotted on the $\phi = \pi / 2$ cut as a function of θ. For small spheres, the E_x component has a maximum at $\theta = \pi / 2$. As the spherical particle gets larger, we observe a shift of the location at which the E_x component has a maximum field. This is due to the increased interaction between a larger sphere and a wider range of angular components of a focused beam. As the size of the spherical particle gets larger, the particle interacts more with components that are incident to larger angles. A similar shift is also observed in the E_z component, as shown in Fig. 3 (a).

In Fig. 3 (b), various components of the near-field electric field are plotted for a radially polarized incidence beam. The results are plotted for spherical particles with radii 70 and 140 nm. E_x and E_z components are plotted on the $\phi = \pi / 2$ cut as a function of θ. The incident beam parameters are identical to the previous set of results with the exception that a radial polarization is used instead of a linear polarization. Contrary to the results in Fig. 3 (a), E_x shows a minimum at $\theta = \pi / 2$ in Fig. 3 (b). This is due to the difference in the strength of various components of the linearly and radially polarized incident focused beams. For the linearly polarized focused wave, the x-component of the electric field is much stronger than the other two components. The radially polarized wave, on the other hand, has a strong z-component in the focal region.

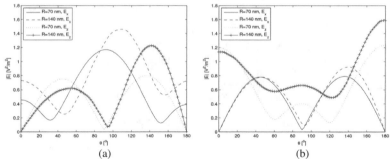

(a)   (b)

Figure 3. Electric field components when a focused beam of light interacts with spheres of various sizes: (a) linearly polarization, (b) radial polarization.

CONCLUSIONS

In summary, an integral equation based numerical solution was developed. The formulations for both plane waves and focused beams were given. For focused beams, the Richards-Wolf vector field equations were combined with the integral equation based formulation to model both linearly and radially polarized focused beams. The results of the integral equation based solution were compared to the results of the analytical Mie series solution for spherical particles and the finite element method for rectangular metallic particles. The methods showed a good agreement.

REFERENCES

1. T. D. Milster, "Horizons for optical data storage," Optics and Photonics News **16**, 28–32 (2005).
2. R. E. Rottmayer, S. Batra, D. Buechel, W. A. Challener, J. Hohlfeld, Y. Kubota, L. Li, B. Lu, C. Mihalcea, K. Mountfield, K. Pelhos, C. Peng, T. Rausch, M. A. Seigler, D. Weller, and X. Yang, IEEE Trans. Magn. **42**, 2417–2421 (2006).
3. A. Hartschuh, E. J. Sanchez, X. S. Xie, and L. Novonty, Phys. Rev. Lett. **90**, 095503 (2003).
4. L. Wang and X. Xu, J. Microsc. **229**, 483–489 (2008).
5. B. Liedberg, C. Nylander, I. Lundstroem, Sens. Actuators **4**, 299–304 (1983).
6. W. A. Challener, I. K. Sendur, and C. Peng, Opt. Express **11**, 3160–3170 (2003).
7. J. T. II Krug, E. J. Sánchez, and X. S. Xie, J. Chem. Phys. **116**, 10895 (2002).
8. T. Yamaguchi, Electron. Lett. **44**, 4455427 (2008).
9. T. Yamaguchi and T. Hinata, Opt. Express **15**, 11481-11491 (2007).
10. L. Liu and S. He, Appl. Opt. **44**, 3429-3437 (2005).
11. T. Grosges, A. Vial, and D. Barchiesi, Opt. Express **13**, 8483-8497 (2005).
12. K. Sendur, W. Challener, and C. Peng, J. Appl. Phys. **96**, 2743–2752 (2004).
13. J. P. Kottmann and O. J. F. Martin, Opt. Express **8**, 655-663 (2001).
14. J. P. Kottmann and O. J. F. Martin, IEEE Trans. Antennas Propag. **48**, 1719-1726 (2000).
15. J. P. Kottmann, O. J. F. Martin, D. R. Smith, and S. Schultz, Chem. Phys. Lett. **341**, 1-6 (2001).
16. J. P. Kottmann, O. J. F. Martin, D. R. Smith, and S. Schultz, New J. Phys. **2**, 27 (2000).
17. J. Jung and T. Sondergaard, Phys. Rev. B **77**, 245310 (2008).
18. F. J. Garcia de Abajo and A. Howie, Phys. Rev. B **65**, 115418 (2002).
19. V. Myroshnychenko et al., Adv. Mater. **20**, 4288-4293 (2008).
20. V. Myroshnychenko et al., Chem. Soc. Rev. **37**, 1792-1805 (2008).
21. J. H. Richmond, Proc. IEEE **53**, 796-804 (1965).
22. R. F. Harrington, Proc. IEEE **55**, 136-149 (1967).
23. R. F. Harrington, *Field Computation by Moment Methods*, (IEEE Press, New York, NY, 1993).
24. E. K. Miller, L. Medgyesi-Mitschang, and E. H. Newman, Eds., *Computational Electromagnetics* (IEEE Press, New York, NY, 1992).
25. R. C. Hansen, Ed., *Moment Methods in Antennas and Scattering*, (Artech, Boston, MA, 1990).
26. A. W. Glisson and D. R. Wilton, IEEE Trans. Antennas Propag. **28**, 593-603 (1982).
27. S. M. Rao, D. R. Wilton, and A. W. Glisson, IEEE Trans. Antennas Propag. **30**, 409-418 (1982).
28. E. Wolf, Philos. Trans. R. Soc. London Ser. A **253**, 349-357 (1959).
29. B. Richards and E. Wolf, Philos. Trans. R. Soc. London Ser. A **253**, 358-379 (1959).

Metamaterials

Mater. Res. Soc. Symp. Proc. Vol. 1182 © 2009 Materials Research Society   1182-EE15-02

Large area light propagation in quasi-zero average refractive index materials

Principia Dardano[1], Vito Mocella[1], Stefano Cabrini[2], Allan. S. Chang[2], Luigi Moretti[1], Ivo Rendina[1], Deindre Olynick[2], Bruce Harteneck[2], and Scott Dhuey[2]
[1]CNR-IMM Unità di Napoli, Via P. Castellino 111, 80131 Napoli, Italy
[2]Molecular Foundry, Lawrence Berkeley National Laboratory, Berkeley, California, USA

ABSTRACT

In this paper the experimental results show near-infrared light collimation through large area (2 x 2 mm) nanopatterned material with refractive index quasi-zero on the average. This quasi- zero refractive index is obtained alternating photonic crystals strips with effective refractive index n_{eff} = −1 and air strips (n = 1). Layers optically annihilate each other, verifying the optical antimatter concept theoretically proposed by Pendry et al [J. Phys.: Condens. Matter 15, 6345 (2003)].

INTRODUCTION

New possibilities [1–4] on electromagnetic waves controlling have been reached using Negative Index Materials (NIMs) [5] which properties result unavailable for natural positive-index materials. In particular the use of NIMs in imaging [6] and magnifying [7] for superlens applications has been object of recent experimental and theoretical studies. Nevertheless , the self-collimation [8,9] of a beam light propagating in NIMs has remained almost unexplored, although this effect can overcome the common diffraction-induced beam spreading. This lack is simply explained by the difficult to long controlling light propagation. Negatively refracting dielectric photonic crystals (PhCs) [10,11] represent a viable route for realizing NIMs capable to overcome this difficulty thanks to their low absorption losses.
Recently, the amplification of evanescent waves in PhCs was verify by experimental evidence of subdiffraction imaging in the near-infrared using negatively refracting PhCs [12].
Besides, PhCs, showing negative refraction behaviour, can accept and capture evanescent light at all angles preserving subwavelength beam size and transferring near-field information.

THEORY

Consider a 2D PhC made of air holes in silicon arranged in a hexagonal lattice. To provide at vertical light confinement the structure was patterned in 1.5 μm thick silicon layer on a 1 μm thick oxide layer of a silicon-on-insulator (SOI) wafer. In this case the index contrast at interfaces and the silicon thickness are so large that the structure behave as a ideal 2D system. Indeed, the effective index of the fundamental mode supported by a uniform slab of this thickness is indistinguishable from the bulk silicon value for both polarizations (n_{eff} = 3.45 for λ = 1.55 μm). The so arranged PhC shows an effective index n_{eff} = −1 for TM polarization (the electric field directed along the holes axis) when the ratio of hole radius r and lattice parameter a is r/a = 0.38 at the normalized frequency ω_n = 0.305. For λ = 1.55 μm, this corresponds to r = 180 nm and a = 472 nm. The main tool to investigate the proprieties of negative refraction behaviour of PhCs is the analysis of EquiFrequency Surface (EFS) that represents the locus in the reciprocal space of propagating wave vectors supported by the unbounded PhC structure at a fixed frequency. The EFS of a PhC described previously around ω_n expands it self with increasing frequency as in positively refracting media till it collapses to the origin of the Brillouin zone. The EFS theoretically calculated at $\omega_n = 0.305$ has circular shape as an isotropic medium, whereas the group velocity, $\vec{v}_g = \partial\omega/\partial\vec{k}$, is inward directed, *i. e.* in the opposite direction of the corresponding wave vector, so that $\vec{v}_g \cdot \vec{k} < 0$. Such a PhC acts as an effective negative refractive index medium [10,11].

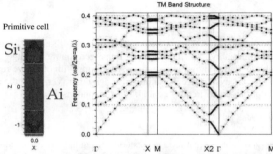

Figure 1: Computed band structure of zero average refractive index material based on effective negative refractive index photonic crystal. Calculation are based on the primitive cell displayed on the left of figure, composed of equal lengths of negative refracting PhC and air. The zero-average bandgap appears at ω_n =0.305.

By alternating several pair of striped layers of this PhC with striped layers of air of equal length, a zero averaged refractive index is obtained. Each pair of PhC layer and air layer can be regarded conceptually as a pair of complementary media (in the real space) that optically cancel each other out, resulting non-existent for the incident light at $\omega_n = 0.305$ [13–15]. In the limit of a very small layer length, the negative refraction of the alternating layers can be utilized to collimate a light beam travelling through the metamaterial system [16,17]. In Figure 1(b) the periodic supercell structure is depicted; it is obtained by stacking three rows of the hexagonal PhC (i.e. a length of $\sqrt{3}\ a$ corresponding to one unit cell of the PhC in the z-direction), with an equal length of air. Nevertheless, when an exactly zero average refractive index is obtained by alternating positive and negative index materials, a full photonic bandgap appears [18, 19]. This is highlighted in the band structure (Fig. 1c) calculated by using the supercell (Fig. 1b) with plane wave expansion (PWE) method. Summing up, at the normalized frequency ω_n =0.305, exactly where n_{eff} (PhC) = −1, the zero-average index condition is met but a very narrow photonic bandgap also appears. Simulation of propagation in this structure using a Finite Difference Time Domain (FDTD) code confirms that a incident beam at the normalized frequency is completely reflected by the structure at this frequency, in agreement with a zero-average-index bandgap.

So, to open up a propagating state through the structure and then improve light transmission, the PhC with negative refraction behavior and air cannot be exactly complementary. Therefore we create a slight imbalance between the positive and negative components of the heterostructure by slightly modifying the width of each PhC strip, but largely preserving the complementary nature of the PhC/air pairs. Thus, the zero-average-index condition is modified to close the bandgap and allow the propagation of light. In the reflection [20] and the subwavelength imaging [21–23] properties of negatively refracting PhCs, the interface termination between PhC and air plays a fundamental role for surface localised states at PhC boundaries. The shape of surface termination can change the surface impedance of the PhC till to it matches the energy flux at the PhC interface and so it induces bangdap closing. By managing the PhC termination and strip width slightly to move away from the zero average index condition, the transmission of the incident wave is maximized and the coupling of the evanescent component is optimized.

This optimization requires a Quasi-Zero Average Refractive Index (QZARI) structure where each PhC stripe is $(3\sqrt{3}-0.4)a$ wide, corresponding to $0.92a$ in air. The Fig. 2 illustrate the light beam propagation scheme (b) and intensity profile propagation computation (c) along the structure. The intensity profile is obtained using 2D FDTD calculations with a Gaussian $\lambda/5$ wide incident beam at a distance of $1.5\sqrt{3}\,a$ from the first PhC stripe and perfectly matched the layer boundary conditions.

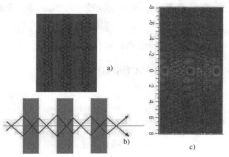

Figure 2 (a) Zero average refractive index material made by alternating layers of air and air holes in silicon photonic crystal arranged in hexagonal lattice (b) Sketch of light propagation in QZARI structure and (c) electric field component E_y profile (TM polarization) inside the QZARI structure calculated using a FDTD code.

Simulation shows that the beam is refocused in every stripe as it propagates, resulting in a strong macroscopic self-collimation effect propagating over the entire length of the simulated structure, *i.e.* 100μm. As refractive index of air-PhC structure isn't exactly zero (but quasi-zero), as the refocusing at each pair of stripes isn't located at the same relative position along the propagation direction and the shape of the beam isn't identical after each refocusing. Nevertheless, an extremely collimated shape is preserved in the absence of any physical waveguide structure that laterally confines the beam along the structure. This effect does not involve any nonlinearity and can in principle be carried on over large distances.

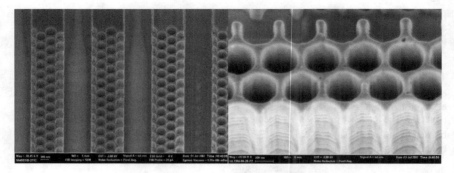

Figure 3: SEM images of the nanofabricated device. (a) Alternating of space and photonic crystal material, with different termination on each PhC structure. (b) Magnified image.

EXPERIMENTAL RESULTS AND DISCUSSIONS

The 2x2 mm device was fabricated on a SOI substrate using a high-definition nanofabrication technology involving high-voltage electron beam lithography and a gas-chopping inductively-coupled plasma etching process [24]. Scanning electron microscope (SEM) images (Fig. 3) of the device shows airholes with a diameter of 360 nm and periodicity of 470 nm with high uniformity overall the PhC stripe. Fig. 4 is a sketch of the experimental setup for the optical characterization of the device. A CW laser provide for light source at λ =1.55µm and is connected to a lensed input fiber producing an incident full-width-half-maximum (FWHM) beam spot size 2.9µm wide at focal distance of 13 µm and experimentally measured. The input and output fibers are put on two different 3-axis nanopositioning stages (NanoMax by Thorlabs) with a resolution of 20 nm (in piezo-electrical retroaction mode) and they are used to optimize coupling between the incident beam rising from the lensed fiber and the QZARI structure. Finally, an IR camera Xenics Xeva 185 and a high numerical aperture (NA=0.42) objective with a long working distance was used to directly observe the light propagation inside the structure. To monitor the light propagation, scattering out-of-plane (x,z) from the QZARI structure is achieved by moving the lensed fiber in the y-direction. This produces a slight misalignment between the input fiber and silicon guiding layer in the SOI, resulting that the incident wavevectors in air will have a small non null vertical component.

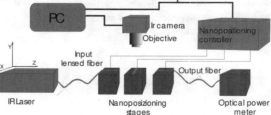

Figure 4: Experimental set-up for transmition analysis

The overall QZARI structure exhibits a periodic modulation of refractive index in the z-direction like in a Bragg grating due to the alternating air and PhC strips (Fig. 5a). The period Λ of such an effective grating profile is given by the distance between two air/PhC pair: $\Lambda = 3\sqrt{3}\,a$ =2.453µm. The radiation scattered out of plane (x,z) will couple with the mth-order of this grating, if the z-component in (y,z) plane of scattered light couple with the grating periodicity. In particular if $k_in = 2\pi/\lambda$ is the incident wavevector component in the (x,z) plane, the z-component in (y,z) plane will be $k_{z,m} = k_in + m2\pi/\Lambda$. This is equivalent to the application of the grating equation $m\lambda = \Lambda(\sin\alpha+\sin\beta)$, where the incident wavevector k_in forms an angle $\alpha = -90°$ with the normal and consequently the mth-orders of diffraction, have angle β : $\sin\beta = (1 + m)\,\lambda\Lambda$ where m = ±1, ±2... with the normal. An experimental angular scan, performed in the (y,z) plane (Fig. 5b), clearly shows the diffraction peak located exactly at the angle $\beta = \pm 21.5°$ corresponding to m = ±1. The angular scan also proves that only a small fraction of energy is coupled to m = −1 grating order, whereas the propagating wave, for which $\beta = 90°$ and m = 0, is two orders of magnitude larger. So, IR images of the light beam was performed simply rotate at $\beta_{-1} = 21.5°$ the IR CCD camera to maximize the light scattered out of plane (x, z) signal.

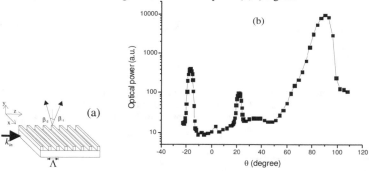

FIG. 5: Grating coupling and angular scan data: (a) Out-of-plane scattering couples with grating direction β_{-1},β_{-2}. (b) Normalized optical power of light scattering in the (y,z) plane in logarithmic scale vs. angle of detection.

The surface termination design minimizes the losses due to impedance and mode mismatch in the coupling of the input lensed fiber with the structure. These insertion losses was theoretical estimated to be less than 0.2 dB, whereas experimentally we found they are at least 1.2 dB. IR images recorded provide clearly the collimation of propagating beam along the entire QZARI structure (Fig. 6a). This result is highlighted by comparing to equivalent propagation over unpatterned SOI (Fig. 6b) recorded with an input power two orders of magnitude larger. The beam confinement and preservation of beam width in the x direction along the entire propagation distance, without any structure to laterally limit the beam in this direction, is proved by the beam profiles (relative intensity vs. x position) shown in Figs. 6(c-e). From these experimental data we estimate a FWHM =2.9 ± 0.4µm of beam propagating in the first part of the sample (Fig. 6(c)); FWHM =3.2±0.4µm in the middle part (Fig. 6(d)) and FWHM =3.0 ± 0.4µ at the end of the sample (Fig. 5(e)). In any case, the FWHM of the transmitted beam is compatible, within the experimental errors, with the measured FWHM of the incident beam emerging from input lensed fiber.

This proves beam spreading is negligible over the entire length of QZARI structure. Besides, in our experiments the beam intensity is largely preserved throughout the device, suggesting vertical scattering losses from the device are low. Furthermore, the experimental results are consistent with the theoretical predictions.

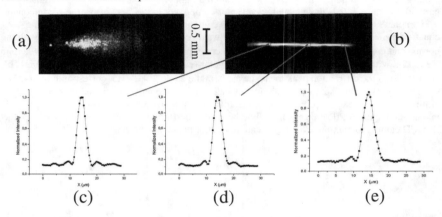

FIG. 6: (a)Experimental images of the scattered radiation along the β_{-1} direction over the QZARI structure, the image reveals a well-collimated beam along the whole 2 mm length of the sample.(b-d) beam profile of the scattered radiation in the first part, (b) in the middle and (d) in the final part of the QZARI structure.

A further analysis of the QZARI structure optical properties of was introduced by measuring the propagation depth on the sample (which total length is 2mm) by varying the incident wavelength and consequently the incident wavevector $\vec{k}_{in}$. Using a tunable laser with $\lambda = (1520 \div 1620)$ nm in the same setup in figure 4, we estimate the propagation length from magnified IR images recorded. We can't appreciate, within the entire length of sample and the experimental errors, a meaningful variation of propagation length around $\lambda = 1550$ nm. Conversely in the range between $\lambda = (1690 \div 1620)$ nm, the length decrease of about 50%. This drastic reduction in the propagation length is explained by the incoming approach to the gap wavelength range of zero average. Indeed, by changing the wavelength, the incident light experience a different effective refractive index of the PhC, $i.\ e.\ n_{eff} = n_{eff}(\lambda)$. So, n_{eff} slightly decrease with the wavelength and the optical path in the PhC slightly balance the optical path in air. In this way, reappear the zero average condition and so the propagation forbidden gap.

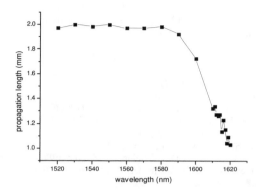

FIG. 7: Propagation length estimated from experimental images of the scattered radiation over the nanopatterned structure vs. wavelength. The propagation length strongly decreases in approaching the gap wavelength typical of zero average refractive index materials.

CONCLUSIONS

Since the refractive index of a PhCs can be considered negative just in the sense of an effective refractive index description [10, 11], and since the reconstruction of all order evanescent components of incident beam is limited by the real finite size of the PhC structure (in comparison to wavelength), it is clear that this structure isn't an ideal complementary media as described in [13]. So, arbitrarily small details of the source at the image plane is limited by the impedance mismatch between air and PhC [25]. Nevertheless, taking into account such limitations, simulation shows sub-wavelength spot size can be obtained over a large scale, whereas experiments already verifying the amplification of evanescent components by a single slab of very similar PhC metamaterial [12]. These results provide for the possibility of supercollimation at subwavelength beam spot size and efficient long-range spatial transfer of near-field information. They open the road to a wide range of applications unachievable using negative-only or positive-only media. For example, functional elements could be incorporated in the positive index portion of the metamaterial; the negative and positive portions can be individually designed and optimized, adding extra degrees of freedom to the design.

In the main, the results presented here is a first experimental verify of the concept of optical antimatter [13]. A strip of PhC appears to annihilate an air strip of equal width at the frequency for which $n_{eff} = -1$. The global effect is as if a 2 mm space was optically removed becoming invisible to the specific wavelength.

This work was shared with the Molecular Foundry, Lawrence Berkeley National Laboratory, under Contract No. DE-AC0205CH11231 supported by the Office of Science, Office of Basic Energy Sciences, of the U.S. Department of Energy.

REFERENCES

1. J. Pendry, Phys. Rev. Lett. 85, 3966 (2000).
2. U. Leonhardt, Science 312, 1777 (2006).
3. J. Pendry, D. Schurig, and D. Smith, Science 312, 1780 (2006).
4. K. Tsakamakidis, A. Boardman, and O. Hess, Nature 450, 397 (2007).
5. V. Shalaev, Nature Phot. 1, 41 (2007).
6. N. Fang, H. Lee, C. Sun, and X. Zhang, Science 308, 534 (2005).
7. Smolyaninov, Y.-J. Hung, and C. Davis, Science 315, 1699 (2007).
8. P. Rakich, M. Dahlem, S. Tandon, M. Ibanescu, M. Soljac, G. Petrich, J. Joannopoulos, L. A. Kolodziejski, and E. P. Ippen, Nature Mat. 5, 93 (2006).
9. Z. Lu, S. Shi, J. Murakowski, G. Schneider, C. Schuetz, and D. Prather, Phys. Rev. Lett. 96, 173902 (2006).
10. M. Notomi, Phys. Rev. B 62, 10696 (2000).
11. M. Notomi, Optical and Quantum Electronics 34, 133 (2002).
12. R. Chatterjee, N. Panoiu, K. Liu, Z. Dios, M. Yu, M. Doan, L. Kaufman, R. Osgood, and C. Wong, Phys. Rev. Lett. 100, 187401 (2008).
13. J. Pendry and S. Ramakrishna, J. Phys.: Condens. Matter 15, 6345 (2003).
14. J. Pendry, Contemp. Phys. 45, 191 (2004).
15. S. Ramakrishna, Rep. Prog. Phys. 68, 449 (2005).
16. E. Shamonina, V. Kalinin, K. Ringhofer, and L. Solymar, Electron. Lett. 37, 1243 (2001).
17. S. A. Ramakrishna, J. Pendry, M. Wiltshire, and W. Stewart, J. Mod. Opt. 50, 1419 (2003).
18. J. Li, L. Zhou, C. Chan, and P. Sheng, Phys. Rev. Lett. 90, 083901 (2003).
19. N. Panoiu, R. Osgood, S. Zhang, and S. Brueck, J. Opt. Soc. Am. B 23, 506 (2006).
20. V. Mocella, P. Dardano, L. Moretti, and I. Rendina, Opt. Express 15, 6605 (2007).
21. S. Xiao, M. Qiu, Z. Ruan, and S. He, Appl. Phys. Lett. 85, 4269 (2004).
22. X. Zhang, Phys. Rev. B 71, 165116 (2005).
23. A. Martinez and J. Marti, Phys. Rev. B 71, 235115 (2005).
24. K. Webb and M. Yang, Phys. Rev. E 74, 016601 (2006).
25. H. Schouten, T. Visser, D. Lenstra, and H. Blok, Phys. Rev. E 67, 036608 (2003).
26. D. Olynick, A. Liddle, and I. Rangelow, J. Vac. Sci. Technol. B 23, 2073 (2005).
27. S. A. Ramakrishna, S. Guenneau, S. Enoch, G. Tayeb, and B. Gralak, Phys. Rev. A 75, 063830 (2007).

Optical Nanoantennas and
Decay Engineering

Mater. Res. Soc. Symp. Proc. Vol. 1182 © 2009 Materials Research Society   1182-EE16-03

Near-Field Radiation From Nano-Particles and Nano-Antennas Illuminated With a Focused Beam of Light

Kursat Sendur, Ahmet Sahinoz, Eren Unlu, Serkan Yazici, and Mert Gulhan
Sabanci University, Istanbul, 34956, Turkey

ABSTRACT

The interaction of photons with metallic nanoparticles and nanoantennas yields large enhancement and tight localization of electromagnetic fields in the vicinity of nanoparticles. In the first part of this study, the interaction of a spherical nanoparticle with focused beams of various angular spectra is investigated. This study demonstrates that the focused light can be utilized to manipulate the near-field radiation around nanoparticles. In the second part of this study, the interaction between linearly and radially polarized focused light with prolate spheroidal nanoparticles and nano-antennas is investigated. Strong and tightly localized longitudinal components of a radially polarized focused beam can excite strong plasmon modes on elongated nanoparticles such as prolate spheroids. The effect of a focused beam on parameters such as the numerical aperture of a beam and the wavelength of incident light, as well as particle geometry and composition are also studied.

INTRODUCTION

The interaction of photons with metallic nanoparticles and nanoantennas is important to a number of emerging nanotechnology applications due to the large enhancement and tight localization of electromagnetic fields in the vicinity of nanoparticles and nanoantennas. This interaction has potential applications at the nanoscale, including near-field scanning optical microscopy [1], optical [2] and magneto-optical [3] high-density data storage, nano-lithography [4], bio-chemical sensing [5], nanoparticle-tweezers [6], and plasmonic solar cells [7]. The plasmon resonance of metallic nanoparticles is a well-studied field [8-9]. The effects of the wavelength, the surrounding medium, the composition, and the shape of the nanoparticle have been investigated in detail [10-11].

Although experimental studies in the literature have used both collimated and focused beams to excite surface plasmons [12-17], until recently the analytical and numerical models in the literature have only used simple plane waves to analyze this interaction. Recently, there has been increasing interest in understanding the interaction of a focused beam of light with a nanoparticle using both numerical and analytical techniques. Numerical techniques based on finite difference time-domain [18] and finite element method [19], as well as analytical techniques based on generalized Mie theory have been used [19-23] to analyze the interaction of a focused beam with a nanoparticle. Focused beam models have also been utilized for other nanostructures, such as nanowaveguides for potential utilization in high density data storage [24]. In a more recent study Mojarad et al. [25] used a radially polarized focused beam to tailor the localized surface plasmon spectra of nanoparticles.

In this study, we investigate the effect of the angular spectrum of a focused beam of light on the near-field radiation from spherical nanoparticles, prolate spheroidal nanoparticles, and dipole nano antennas. Focused beams with various angular specta are utilized in this study.

SPHERICAL NANOPARTICLES

To analyze the effect of the angular spectrum on the near-field radiation from a spherical nanoparticle, a silver nanoparticle is illuminated using a focused beam of light with small and large α. The focused beam propagates in the z-direction and is polarized in the x-direction. In Fig. 1, the electric field is computed at various wavelengths on the x-z cut-plane for a silver sphere with a 250 nm radius. The field distributions in Fig. 1 are normalized to the value of the incident intensity at the focus. At each wavelength, the field distribution $E_x(x,0,z)$ and $E_z(x,0,z)$ is plotted for $\alpha = 5°$ and $\alpha = 60°$. A comparison of Figs. 1 (a) and (d) suggests that the field distribution at $\lambda = 400$ nm for $\alpha = 5°$ shows a significant difference compared to the results of $\alpha = 60°$. In Figs. 1 (a)-(f), deviations are observed at other wavelengths as well. The E_y component is negligible for the solutions. The impact of altering the angular spectrum is more drastic for the E_z component, as shown in Figs. 1 (g)-(l). For example, when the angular spectrum is narrowly distributed along the direction of propagation, as shown in Fig. 1 (i), two stronger lobes are observed at the back of the spherical particle. As α is increased, and therefore the angular spectrum is widened, the stronger lobes are moved from the back of the particle to the front of the particle, as shown in Fig. 1 (l). This was achieved without changing the frequency, geometry, or composition of the particle. Suppressing strong near-field radiation lobes and enhancing weaker radiation lobes is possible by altering the angular spectrum.

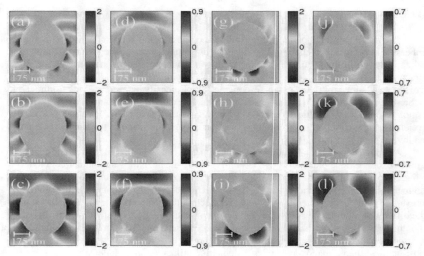

Figure 1. $E_x(x,0,z)$ and $E_z(x,0,z)$ on x-z for various $[\alpha,\lambda]$: (a) $E_x(x,0,z)$ for [5,400], (b) $E_x(x,0,z)$ for [5,500], (c) $E_x(x,0,z)$ for [5,600], (d) $E_x(x,0,z)$ for [60,400], (e) $E_x(x,0,z)$ for [60,500], (f) $E_x(x,0,z)$ for [60,600], (g) $E_z(x,0,z)$ for [5,400], (h) $E_z(x,0,z)$ for [5,500], (i) $E_z(x,0,z)$ for [5,600], (j) $E_z(x,0,z)$ for [60,400], (k) $E_z(x,0,z)$ for [60,500], and (l) $E_z(x,0,z)$ for [60,600],

PROLATE SPHEROIDAL NANOPARTICLES

In this section, the impact of the angular spectrum distribution of the incident radially polarized beam on the near-field radiation of spheroidal nanoparticles is studied. In Fig. 2 (a), a schematic illustration of a prolate spheriodal nanoparticle and the incident radially polarized focused beam is provided. In Fig. 2 (b), the total intensity profile is plotted on the x-z plane, which passes through the center of a gold prolate spheroid particle with a major/minor axis ratio of 5. In Fig. 3, $|E|^2$ distributions for a gold prolate spheroid are given for various half-beam angles. In this figure, the prolate spheroids are illuminated with a radially focused beam with half-beam angles $\alpha=15°$, $\alpha=30°$, $\alpha=45°$, and $\alpha=60°$. The field distributions in Figs. 2 and 3 are normalized to the value of the incident intensity at the focus. The results suggest that the electric field distribution does not change as the half-beam angle is increased. The amplitude of the near-field electric field distribution, however, increases as the half-beam angle is increased. The angular spectrum of the incident beam is tight for $\alpha=15°$, becoming wider as the half beam angle is increased. Therefore, the incident wave amplitude onto the particle increases as the half-beam angle increases. As a result of increasing the incident field amplitude, the scattered field amplitude also increases, as shown in Fig. 3.

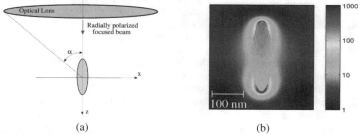

(a)   (b)

Figure 2. (a) A schematic illustration of a prolate spheriodal nanoparticle and the incident radially polarized focused beam. (b) $|E|^2$ distribution on the x-z cut plane for a gold prolate spheroid particle of major axis radius of 100 nm and a major/minor axis ratio of 5 illuminated with a radially polarized focused light at $\lambda =700$ nm.

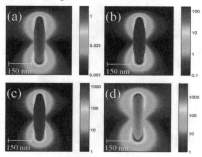

Figure 3. $|E|^2$ distributions for a gold prolate spheroid. The distributions are given for various half-beam angles: (a) $\alpha=15°$, (b) $\alpha=30°$, (c) $\alpha=45°$, and (d) $\alpha=60°$.

NANOANTENNAS

Figure 4 illustrates the focused incident $|E^i|^2$ distributions onto the dipole nano-antennas. The fields are plotted at the focal plane x-y for various half-beam angles. For small α, the field distribution is similar to that of a plane wave. As the α increases, the beam becomes more tightly focused. In Fig. 5, the total $|E^t|^2$ distribution for a dipole antenna is shown on the x-z cut-plane plane when it is illuminated with the focused beams shown in Fig. 4. For this particular simulation, the sizes of the antenna are L= 110, T= 20, W= 20, and G= 20 nm. In this simulation, the incident focused beam is polarized in the x-direction, and propagates in the negative z-direction. The wavelength of the incident light is 850 nm. The antenna is placed at the focus of the incident beam, which is also the global origin for the simulations. The peak value of the graphs in Fig. 5 represents the intensity enhancement, which is defined as

$$\text{Intensity Enhancement} = \frac{\left|E^t(0,0,0)\right|^2}{\left|E^i(0,0,0)\right|^2} \tag{1}$$

In other words, total intensity $|E^t|^2$ at the antenna gap center is normalized to the value of the incident intensity $|E^i|^2$ at the focus. The results in Fig. 5 show a confined electric field close to the gap region of the antenna. Also, a large electric field enhancement is observed for all the half-beam angle values α. The intensity enhancement at the center of the antenna is about 1200 for small α. The intensity enhancement slowly reduces to 1100 as α is increased.

The results in Fig. 5 indicate that the contribution from the rays with large incident angles is less than the contribution from the rays with small incident angles. To provide further evidence for this observation, the incident beam is separated into angular spectral bands. The contribution from each spectral band is then calculated separately. To achieve this, the incident beam is passed through an angular band-pass filter as shown in Fig. 6. The filter supresses the $\theta<\theta_{min}$ and $\theta>\theta_{max}$ part of the angular spectrum. In this calculation, spectral bands with 5° intervals are considered, which corresponds to $\theta_{min} = \theta_{min+5}°$. Intensity enhancement is plotted as a function of θ_{min} in Fig. 7. The results suggest that the intensity enhancement due to rays with large incident angles is less than the intensity enhancement from the rays with small incident angles.

As we mentioned above, the results in Fig. 5 indicates that the intensity enhancement is almost preserved as α increases. The output power, however, not only depends on the intensity enhancement but also on the intensity of the incident focused beam. As shown in Fig. 8, the intensity of the incident focused beam $|E^i(0,0,0)|^2$ increases with increasing α. The incident beam becomes more tightly focused with increasing α, which increases the incident electric field intensity, as shown in Fig. 8. As a result, the output intensity increases with increasing α. This suggests that significant gains can be achieved by increasing the α, despite a small reduction in the transmission efficiency.

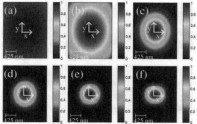

Figure 4. Incident $|E|^2$ distribution in the absence of nanoantennas. The fields are plotted at the focal plane x-y for: (a) α =5°, (b) α =15°, (c) α =30°, (d) α =45°, (e) α =60°, and (f) α =75°.

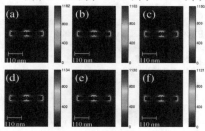

Figure 5. Total $|E|^2$ distribution in the presence of nano-antennas. The fields are plotted on the x-y cut plane for: (a) α =5°, (b) α =15°, (c) α =30°, (d) α =45°, (e) α =60°, and (f) α =75°.

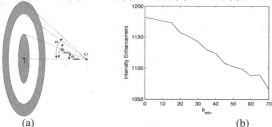

(a)   (b)

Figure 6. (a) A band-pass filter supresses the $\theta<\theta_{min}$ and $\theta>\theta_{max}$ part of the angular spectrum. (b) Intensity enhancement as a function of θ_{min}. The cut-off angle is selected as $\theta_{min} = \theta_{min+}5°$.

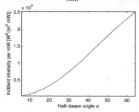

Figure 7. Incident intensity per mW incident power as a function of α for λ=850 nm.

111

CONCLUSIONS

In summary, it has been demonstrated that the near-field radiation from a spherical particle can be manipulated by adjusting the angular spectrum of an incident focused beam. On the other hand, for prolate spheroids the electric field distribution does not change as the half-beam angle is increased. The amplitude of the near-field electric field distribution, however, increases as the half-beam angle is increased. For nano-antennas the intensity enhancement is almost preserved as α increases. For more tightly focused beams, the output intensity increases with increasing α. This suggests that for nano-antennas, significant gains can be achieved by increasing the α despite a small reduction in the transmission efficiency.

REFERENCES

1. A. Hartschuh, E. J. Sanchez, X. S. Xie, and L. Novonty, Phys. Rev. Lett. **90**, 095503 (2003).
2. T. D. Milster, Optics and Photonics News **16**, 28–32 (2005).
3. R. E. Rottmayer, S. Batra, D. Buechel, W. A. Challener, J. Hohlfeld, Y. Kubota, L. Li, B. Lu, C. Mihalcea, K. Mountfield, K. Pelhos, C. Peng, T. Rausch, M. A. Seigler, D. Weller, and X. Yang, IEEE Trans. Magn. **42**, 2417–2421 (2006).
4. L. Wang and X. Xu, J. Microsc. **229**, 483–489 (2008).
5. B. Liedberg, C. Nylander, I. Lundstroem, Sens. Actuators **4**, 299–304 (1983).
6. Q. Zhan, Opt. Express **12**, 3377–3382 (2004).
7. K. R. Catchpole and A. Polman, Opt. Express **16**, 21793 (2008).
8. J. Crowell and R. H. Ritchie, Phys. Rev. **172**, 436–440 (1968).
9. M. Kerker, Appl. Optics **18**, 1180–1189 (1979).
10. K. L. Kelly, E. Coronado, L. L. Zhao, and G. C. Schatz, J. of Phys. Chem. B **107**, 668–677 (2003).
11. O. Sqalli, I. Utke, P. Hoffmann, and F. Marquis-Weible, J. Appl. Phys. **92**, 1078–1083 (2002).
12. H. Kano, D. Nomura, and H. Shibuya, Appl. Opt. **43**, 2409–2411 (2004).
13. H. Kano, S. Mizuguchi, and S. Kawata, J. Opt. Soc. Am. B **15**, 1381–1386 (1998).
14. A. Bouhelier, F. Ignatovich, A. Bruyant, C. Huang, G. Colas des Francs, J.-C. Weeber, A. Dereux, G. P. Wiederrecht, and L. Novotny, Opt. Lett. **32**, 2535–2537 (2007).
15. A. V. Failla, H. Qian, H. Qian, A. Hartschuh, and A. J. Meixner, Nano Lett. **6**, 1374–1378 (2006).
16. A. V. Failla, S. Jager, T. Zuchner, M. Steiner, and A. J. Meixner, Opt. Express **15**, 8532–8542 (2007).
17. T. Zuchner, A. V. Failla, A. Hartschuh, and A. J. Meixner, J. Microsc. **229**, 337–343 (2007).
18. W. A. Challener, I. K. Sendur, and C. Peng, Opt. Express **11**, 3160–3170 (2003).
19. K. Sendur, W. Challener, and O. Mryasov, Opt. Express **16**, 2874–2886 (2008).
20. J. Lerme, G. Bachelier, P. Billaud, C. Bonnet, M. Broyer, E. Cottancin, S. Marhaba, and M. Pellarin, J. Opt. Soc. Am. A **25**, 493–514 (2008).
21. N. M. Mojarad, V. Sandoghdar, and M. Agio, J. Opt. Soc. Am. B **25**, 651–658 (2008).
22. N. J. Moore and M. A. Alonso, Opt. Express **16**, 5926–5933 (2008).
23. C. J. R. Sheppard and P. Torok, J. Mod. Opt. **44**, 803–818 (1997).
24. K. Sendur, C. Peng, and W. Challener, Phys. Rev. Lett. **94**, 043901 (2005).
25. N. M. Mojarad and M. Agio, Opt. Express **17**, 117–122 (2008).

Appendix

This Appendix contains papers
from Materials Research Society Symposium Proceedings
Volume **1077E**
Functional Plasmonics and Nanophotonics
S. Maier and S. Kawata, Editors

Mater. Res. Soc. Symp. Proc. Vol. 1077 © 2008 Materials Research Society   1077-L01-02

Propagation of High-Frequency Surface Plasmons on Gold

Robert Edwin Peale[1], Olena Lopatiuk-Tirpak[1], Justin W. Cleary[1], Samantha Santos[1], John Henderson[1], David Clark[1], Leonid Chernyak[1], Thomas Andrew Winningham[1], Enrique Del Barco[1], Helge Heinrich[1], and Walter R. Buchwald[2]

[1]Physics, University of Central Florida, 4000 Central Florida Blvd., Orlando, FL, 32816
[2]RYHC, Air Force Research Lab, 80 Scott Drive, Hanscom AFB, MA, 01731

ABSTRACT

Surface plasmon propagation on gold over 0.1 – 1.6 micrometer distances for plasmon energies in the range 1.6 – 3.5 eV was characterized. Surface plasmons were excited by an electron beam near a grating milled in the gold. The spectra of out-coupled radiation reveal increasingly strong propagation losses as surface plasmon energy increases above 2.8 eV, but little effect in the range 1.6 – 2.8 eV. These results are in partial agreement with theoretical expectations.

INTRODUCTION

Propagation of electromagnetic signals on metal waveguides via highly-confined, bound electro-magnetic waves known as surface plasmon polaritons (SPP) is central to nano-photonics [1, 2]. Plasmon-electronic integrated circuits (PEIC) have been proposed [3], but the usual optical inputs and outputs for PEIC involve bulky optics. Potentially more compact would be electrical SPP generation and detection, e.g. by electron bombardment using nano-tube field emitters. The spectrum of electron-beam excited SPPs on metals is concentrated at high visible and ultraviolet energies. Although the characteristic propagation length generally decreases with energy for free-electron metals, gold specifically has a propagation length that is expected to remain constant at about 0.3 μm above 2.5 eV (see below). Thus, electron-beam excited, high-frequency SPPs on gold are potentially interesting for nano-scale PEIC applications. Use of a scanning electron microscope (SEM) and cathodoluminescence (CL) to study SPP decay for 1-10 μm propagation lengths at energies below 2.3 eV was recently described [4-6]. This paper reports an independent experiment of the same type, but where sub-micron propagation is observed at SPP energies above 2.8 eV.

EXPERIMENT

A nominally 470 nm thick layer of gold was e-beam evaporated onto a 5 nm Cr sticking layer on a polished silicon substrate. This thickness is sufficient that the optical constants are those of bulk gold. Using a 30 keV focused gallium-ion beam (FIB), several 20-line gratings were cut in the gold. The Figure 1 inset presents an FIB micrograph of one of the gratings, where all twenty of the 57 μm-long lines appear in the upper part of the image.

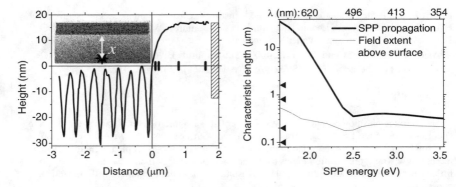

Figure 1. (left) Atomic force microscopy line scan of grating on gold film. The inset is a focused ion beam micrograph of the grating. The black star schematically represents the ~5 nm diameter electron beam spot (exaggerated for clarity) at a distance x (also exaggerated) from the rulings. Cathodoluminescence spectra were collected as a function of x. The bar symbols on the zero line indicate the x-values probed. The shaded box represents the penetration depth of the surface plasmon fields into the metal

Figure 2. (right) Characteristic lengths as a function of SPP energy. The propagation length for SPPs on gold is given by the heavy curve. The light curve is the SPP penetration depth into the air above the surface. The triangle symbols indicate the distances from the grating that SPPs were excited by the electron beam. The tick labeling on the upper border indicates the photon wavelengths for out-coupled SPPs.

     Atomic Force Microscopy (AFM) shows that the as-evaporated surface consists of bumps with 1-4 nm height, 50 nm diameter, and ~50 nm average separation. The Figure 1 AFM image slice reveals 30 nm groove depths, but the profiles are degraded by the AFM tip size [7, 8]. The grating period is 360 nm. Figure 1 also reveals that the regions between the grooves have been collaterally milled and are lower than the surrounding unstructured gold by about 17 nm.

     A scanning electron microscope cathode-luminescence system collects spectra for different electron-beam positions. An off-axis parabolic aluminum mirror collects ~75% of all light emitted from the vicinity of the excitation. The electron beam transits a 1 mm aperture in the mirror to excite SPPs at the mirror focus. Collection system dimensions sufficiently exceed SPP propagation lengths that SPP out-coupling occurs essentially at the focus.

     The Figure 1 inset schematically indicates the e-beam spot at a distance from the grating of x, where surface plasmons with a certain frequency distribution are excited with efficiency < 1% [4, 9]. SPPs propagate away from the excited spot, and those that reach the grating are coupled into free electromagnetic waves with an efficiency that depends on grating geometry [10]. Vertical bar symbols on the zero-line of Figure 1 indicate the actual x-positions probed.

THEORETICAL CONSIDERATIONS

The complex SPP wavevector is determined from the complex permittivity ε of the metal, according to

$$k = (\omega/c)\sqrt{[\varepsilon/(1+\varepsilon)]}, \qquad (1)$$

where ω is the angular frequency and c is the speed of light. Empirical permittivity data for gold [11] were used to obtain the dispersion relation ω vs Re[k]. Between 2 and 2.5 eV (620-500 nm wavelength), the SPP dispersion curve falls noticeably below the light line $\omega = c$ k. Above 2.5 eV, rather than leveling off at an SPP resonance frequency as for free-electron metals, the curve for gold doubles back toward the light line while remaining below it. The fundamental condition for SPPs that Re[ε] < 0 remains valid up to at least 5 eV.

The characteristic propagation length L for SPP intensity is given by $L^{-1} = 2$ Im[k]. Figure 2 presents calculated L values using permittivity from [11] over the experimental spectral range, where we find 0.3 μm < L < 40 μm. Above 2.5 eV (below 500 nm optical wavelength), L is fairly constant at ~ 0.3 – 0.4 μm. Symbols indicate distances x probed in the experiment.

SPP fields extend above the surface and penetrate into the gold by amounts

$$L_{Air} = (c/\omega) / Re[\sqrt{(-1/(1 + \varepsilon))}]$$

and

$$L_{Au} = (c/\omega) / Re[\sqrt{(-\varepsilon^2/(1 + \varepsilon))}], \qquad (2)$$

respectively. L_{Air} values in the experimental spectral range are plotted in Figure 2 and fall between 0.2 and 0.6 μm. Values for L_{Au} are between 25 and 40 nm, and the shaded box in Figure 1 schematically indicates the value at 3.5 eV. Thus, the comparatively small ~17 nm recess of the grating below the surface should not prevent the SPP from feeling the grating, so that out-coupling should proceed in the usual way. However, the spectral efficiency for out-coupling may be very different from that of the gratings in [4-6].

The out-coupling angle θ is determined by

$$Re[k] = (2 \pi/\lambda) \sin \theta + 2 \pi m/a, \qquad (3)$$

where m is a positive integer. The apparatus collects only m = 1 emission over the spectral range of the experiment. For a = 50 nm, the characteristic length scale of surface roughness, there are no real solutions for θ in the experimental spectral range, so that contributions to the background from out-coupling by roughness should be weak.

SPP excitation by electron beams is sharply peaked at the SPP resonance frequency $\omega = \omega_p/\sqrt{2}$ for free-electron metals [12]. Experiment [13] and simulation [14] reveal two peaks in the electron energy loss spectrum for gold due to SPP generation. These occur near 2.8 and 5.7 eV as indicated by symbols in Figure 3. The SPP resonance frequency determined by $\varepsilon = -1$ [15] corresponds to peaks in the electron energy-loss function for SPP generation Im[-1/(ε + 1)] [10], which do occur near 2.8 and 5.7 eV (Figure 3).

117

Next, theoretical considerations relevant to the analysis of the experimental data are presented. The measured emission spectrum $I(x, \lambda)$ is

$$I(x, \lambda) = S(\lambda) \{B(\lambda) + D(\lambda) \, Exp[-x/L(\lambda)]\}. \qquad (4)$$

$D(\lambda)$ is the distribution of surface plasmons at frequency ω excited by the electron beam, which is observed (Figure 3) to fall sharply for energies below 2.5 eV to the level of only a few % of the peak value. The ratio of $I(x, \lambda)$ to $I(x>>L, \lambda)$ is

$$R(x, \lambda) = 1 + [D(\lambda)/B(\lambda)] \, e^{-x/L(\lambda)}. \qquad (5)$$

The second term is small compared with unity, and becomes rapidly smaller with increasing λ due to the factor D/B. $B(\lambda)$, which is determined from $I(x>>L, \lambda)$ and the known spectrometer response function $S(\lambda)$, is usually assumed [4-6] to be independent of x.

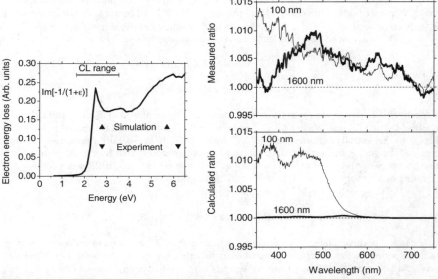

Figure 3. (left) Electron energy loss spectrum for generation of SPPs on gold. The symbols represent experimentally observed [13] and simulated [14] energy loss peaks. The curve is the energy loss function calculated from the permittivity. The range of photon energies collected by the cathodo-luminescence experiment is indicated by the horizontal bar.
Figure 4. (right) Measured (top) and calculated (bottom) ratios of cathodo-luminescence spectra. Distances of electron-beam excitation spot from the grating out-coupler are indicated.

118

RESULTS

Experimental ratios I(x, λ) / I(1 mm, λ) are presented in the upper part of Figure 4 for two distances x. At x = 100 nm, the short-wave emission dominates. At x = 1600 nm, the emission for λ < 450 nm drops to the level of the background, while the longer wave emission remains strong. The lower part of Figure 4 presents calculated ratios Eq. (5) for the same two distances. The calculated ratios adequately reproduce the observed drop in the values for λ < 400 nm. The calculated ratios converge to unity for λ > 700 nm as seen in the data. However, the drop with x between 400 and 700 nm is less rapid, or even non-existent in comparison with the calculated effect.

Figure 5 presents a plot of the experimental ratios for three different wavelengths as a function of distance. The heavy curve is the calculated ratio at 350 nm, and the calculated curves for other wavelengths are similar. At 350 nm, the experimental ratio drops rapidly to unity in fair agreement with the calculation. However, at 450 nm the experimental ratio hardly changes at all, which is significantly different than the calculation.

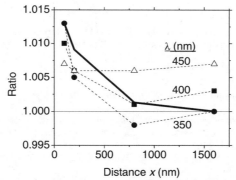

Figure 5. Ratios of cathodo-luminescence spectra at three wavelengths (symbols) as a function of the distance of the electron-beam excitation spot from the grating out-coupler. The heavy curve is the calculated ratio for λ = 350 nm. Calculated curves for other wavelengths are similar.

DISCUSSION

Propagation lengths are usually found to be about 2-7 times smaller than predicted by permittivity-based theory [4-6]. If L is decreased, one finds that the calculated ratio has more of a hump at 500 nm for the larger distances x, but these calculated ratios also collapse toward unity even more rapidly with x. Thus, the disagreement in the middle of the spectral range suggests an oversimplification in the formula, especially in the assumption of position independence of the background spectrum B.

The previous studies describe their background as "significant"[5] or comprising ~75% of the total signal [4]. In our experiment, the differences between spectra near and far from the grating are only about 1%. The origin of the background, and what determines its strength relative to the SPP signal are unclear, though we have argued that SPP out-coupling by surface roughness should be a weak contribution, in contrast to the assumption of [5]. The background may be attributed [4] to d-band emission, dipole radiation originating from incident electrons and their mirror charges, and contaminant fluorescence, but with unknown relative strengths. To explain the weakness of our SPP signal relative to the background, we suppose that our grating is inefficient at outcoupling SPPs in comparison to those of [4-6]. In [4,6] lines of the grating are raised above the gold surface. In [5], the grating was formed by plasma etching of a Si substrate, later coated in Au, and the tops of the grating bars appear to be level with the gold surface. In contrast, our grating was formed by FIB, which caused the entire grating structure to be sunk by 17 nm below the surrounding surface (Figure 1), giving a discontinuity in height between surface and grating.

In summary, propagation of electron-beam excited surface plasmons on gold was characterized over $0.1 - 1.6$ μm distances. The effect of attenuation was observed for surface plasmons with energies in the range $2.8 - 3.5$ eV, but not for the lower energies that had been characterized earlier by others. The results are in partial agreement with theoretical expectations. The disagreement is likely due to simplifying assumptions in the calculation.

ACKNOWLEDGMENTS

The authors wish to acknowledge funding for this work provided by the Air Force Office of Scientific Research Task 06SN05COR and AFRL contract number FA871806C0076.

REFERENCES

1. M. L. Brongersma and P. G. Kik, *Surface Plasmon Nanophotonics* (Springer, New York, 2007).
2. S. A. Maier, *Plasmonics: Fundamentals and Applications* (Springer, New York, 2007).
3. R. A. Soref, in *Silicon Photonics—The State of the Art*, edited by G. Reed (Wiley, Hoboken, NJ, 2008).
4. M. V. Bashevoy, F. Jonsson, A. V. Krasavin, N. I. Zheludev, Y. Chen, and M. I. Stockman, Nano Lett. 6, 1113 (2006).
5. J. T. van Wijngaarden, E. Verhagen, A. Polman, C. E. Ross, H. J. Lezec, and H. A. Atwater, Appl. Phys. Lett. 88, 221111-1 (2006).
6. M. V. Bashevoy, F. Jonsson, K. F. MacDonald, Y. Chen, and N. I. Zheludev, Optics Express 15, 11313 (2007).
7. J. Aue´ and J. Th. M. De Hossona, Appl. Phys. Lett. 71, 1347 (1997).
8. E. C. W. Leung, P. Markiewicz, and M. C. Goh, J. Vac. Sci. Technol. B 15, 181 (1997).
9. H. Raether, in *Physics of Thin Films*, edited by G. Hass, M. H. Francombe, and R. W. Hoffman (Academic, New York, 1977) vol. 9, pp. 145-261.
10. A. V. Krasavin, K. F. MacDonald and N. I. Zheludev, in *Nanophotonics with Surface Plasmons*, edited by V. M. Shalaev and S. Kawata (Elsevier, Amsterdam, 2007), pp. 109-139.
11. P. B. Johnson and R. W. Christy, Phys. Rev. B 6, 4370 (1972).
12. R. H. Ritchie, Phys. Rev. 106, 874 (1957).
13. J. C. Ingram, K. W. Nebesny, and J. E. Pemberton, Appl. Surf. Sci. 44, 293 (1990).
14. Z.-J. Ding and R. Shimizu, Phys. Rev. B 61, 14128 (2000).
15. E. A. Stern and R. A. Ferrell, Phys. Rev. 120, 130 (1960).

Mater. Res. Soc. Symp. Proc. Vol. 1077 © 2008 Materials Research Society   1077-L01-04

All-Optical Active Plasmonics Based on Ordered Au Nanodisk Array Embedded in Photoresponsive Liquid Crystals

Yue Bing Zheng, Vincent K. S. Hsiao, and Tony Jun Huang
Department of Engineering Science and Mechanics, Pennsylvania State University, 212 Earth-Engineering Sciences Building, University Park, PA, 16802

E-mail: junhuang@psu.edu (T. J. Huang)

ABSTRACT

We propose a new approach towards all-optical active plasmonics based on ordered Au nanodisk arrays embedded in azobenzene-doped liquid crystals (LCs). Upon photoirradiation, the doped LCs went through phase transition induced by *trans-cis* photoisomerization of the azobenzene molecules. The phase transition led to the change in the refractive index of the LCs experienced by incident light, and enabled reversible tuning of the localized surface plasmon resonance (LSPR) of the embedded Au nanodisks. The tuning utilized the sensitivity of the LSPR of the Au nanodisks to the change in the surroundings' refractive index. Experimental observations on both peak shift and intensity change of the LSPR matched those from discrete dipole approximation (DDA) calculations.

INTRODUCTION

Plasmonics has both the capacity of photonics and the miniaturization of electronics, offering the potential to merge photonics and electronics at nanoscale dimensions [1]. Plasmonic waveguide, a passive component that acts as interconnector in the integrated circuits, has been intensely studied and shown optimistic for practical applications [2]. However, active plasmonic devices such as switches and modulators have to be developed before the full potential of plasmonics can be realized [3]. For achieving most of the active plasmonic devices, the capability of reversible tuning of the surface plasmon resonance (SPR) of metal nanostructures with control signals is a prerequisite. Chau *et al.* demonstrated that terahertz light transport in spintronic-plasmonic media, a dense ensemble of subwavelength bimetallic ferromagnetic/nonmagnetic microparticles, can be actively controlled by magnetic field [4]. Leroux *et al.* showed that the shape and peak wavelength of LSPR of Au nanoparticles embedded in polyaniline thin films can be reversely tuned by electrochemically changing refractive index of the polyaniline thin films [5]. Electrochemical tuning of LSPR of Ag nanoparticle arrays modified by WO_3 sol-gel was realized by Wang *et al.* [6]. The birefringence in LCs, a large change in refractive index obtained by changing the alignment of LCs, made LCs an ideal functional material for active plasmonics. Active tuning of SPR of different metal nanostructures has been achieved by electrically controlling the alignment of LCs near the surfaces of the nanostructures [7].

Light exhibits several advantages as control signals, including fast response speeds, and simultaneous reading and writing. So does all-optical active plasmonics in which the active tuning of SPR is controlled by light. In this paper, we demonstrate all-optical active plasmonics based on photoresponsive LCs. The photoresponsive LCs were developed through doping nematic LCs with certain amount of azobenzenes. The photoswitchable azobenzene molecules went through *trans-cis* photoisomerization by alternative switching on and off of pump light. The photoisomerization induced phase transition of the LCs due to the cooperative effects among LC molecules, leading to a change in the refractive index of the LCs. The photo-induced change in the LCs caused change in the LSPR of the embedded Au nanodisk array due to the dependence of the LSPR on the surroundings' refractive index (Figure 1).

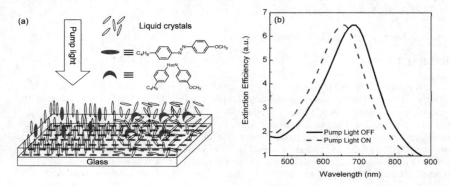

Figure 1. (a)Working principle of all-optical active plasmonics based on Au nanodisk array embedded in photoresponsive LCs. (b) Illustrated LSPR spectra exhibiting the peak shift upon photoirradiation on the LCs by pump light.

EXPERIMENT

The photoresponsive LCs were a homogeneous mixture of 70% nematic LCs (TL213, Merck), 20% 4-butyl-4′-methyl-azobenzene (BMAB), and 10% chiral dopant (S811, Merck). Initially, the doped LCs were thermally treated to be in an isotropic phase with the azobenzenes in the *cis* configuration. Upon photoirradiation, the azobenzenes transformed from *cis* to *trans*, inducing the transition of the LCs from isotropic to nematic phase (Figure 1a).

Ordered Au nanodisk arrays were fabricated on glass substrates by nanosphere lithography (NSL) [8]. Figure 2 shows the schematic of the fabrication procedure. First, an Au thin film of controlled thickness was deposited onto a pre-cleaned glass substrate by a thermal evaporation technique. A thin layer of Cr was sandwiched between the Au film and the glass substrate as an adhesive layer. Second, a self-assembled monolayer (SAM) of close-packed polystyrene (PS) nanospheres was formed the Au surface. Third, O_2 reactive ion etching (RIE) was carried out to morph the close-packed PS nanosphere monolayer into array of separated nanoellipses, and then Ar RIE was followed to selectively etch a portion of the Au and Cr film that was not protected by the nanoellipses. Finally, Au nanodisks were produced on the substrate after the remaining PS was selectively removed by immersing the substrate in Toluene with ultrasonic for 3 min. To enhance the optical properties, the Au nanodisks were annealed in a furnace under ambient atmospheric conditions.

An active plasmonic cell was constructed by sandwiching the photoresponsive LCs between a blank glass slide and the glass substrate with an Au nanodisk array on it. The distance between the glass slide and substrate was defined by the diameter of silica microspheres that were used as spacers.

Scanning electron microscopy (SEM) was used to characterize the topology of the PS nanospheres, PS nanoellipses, and Au nanodisks. Images were taken on a Hitachi S-3500N with an accelerating voltage of 5 kV. Extinction spectra of the Au nanodisk arrays were recorded on an HR4000 spectrophotometer from Ocean Optics Inc. An unpolarized incident light beam was normal to the substrate surface.

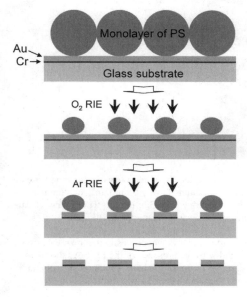

Figure 2. Schematic of the fabrication procedure for ordered Au nanodisk arrays on glass substrate by NSL.

DISCUSSION

Figure 3a shows the SEM image of SAM of PS nanospheres on substrate. The nanospheres have a diameter of 300 nm. The close-packed nanospheres were hexagonally arranged, with triangular interstices formed between every three neighbouring nanospheres. Figure 3b is the tilted-view SEM image of PS nanoellipses after O_2 RIE of PS microspheres with a diameter of 1.05 μm. In O_2 RIE, the parameters included 90 s etching time, 20 sccm (Standard Cubic Centimetres per Minute) gas flow, 100 mTorr pressure, and 300 W power density. Figure 3c shows the SEM image of hexagonally arranged Au nanodisk array on glass substrate. The original PS nanospheres had a diameter of 320 nm. In O_2 RIE, the parameters included 30 s etching time, 20 sccm gas flow, 100 mTorr pressure, and 300 W power density. The same parameters were used in Ar RIE, except that the etching time was

100 s. Figure 3d is the tilted-view SEM image of Au nanodisk array on glass substrate with the PS nanoellipses in Figure 3b as mask during Ar RIE.

Figure 3. SEM images of (a) Original PS nanosphere array. (b) Tilted view of PS nanoellipse array after O_2 RIE on SAM of PS microspheres with a diameter of 1.05 μm. (c) Au nanodisk array produced from SAM of PS nanospheres with a diameter of 320 nm. (d) Tilted view of Au nanodisk array produced from SAM of PS microspheres with a diameter of 1.05 μm.

The effects of LCs on the LSPR of Au nanodisk arrays were investigated by measuring the extinction spectra of the nanodisks without and with the presence of the LCs. As shown in Figure 4a, the solid curve stands for the extinction spectrum of the bare Au nanodisk array, and the dashed curve is the extinction spectrum when the nanodisk array was embedded in the

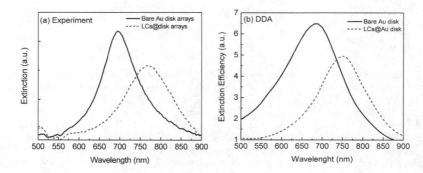

Figure 4. (a) Experimental extinction spectra of Au nanodisk array before (solid curve) and after (dashed curve) LCs. (b) Calculated extinction spectra of Au nanodisk before (solid curve) and after (dashed curve) LCs.

azobenzene-doped LCs. For each extinction spectrum, the single band revealed the LSPR that arose from the in-plane dipole resonance of the Au nanodisks. The LCs caused a redshift from 680 to 748 nm in the LSPR peak with decreased intensity.

DDA calculations of the effects of LCs on the LSPR of Au nanodisks (Figure 4b) yielded good match with the experimental observations. We used a multi-layer model in the open program DDSCAT by Draine and Flatau (version 6.1) to calculate the extinction spectra of the Au nanodisks before and after the LCs were applied [8]. The redshifts of the LSPR peak in both experiments (Figure 4a) and calculations (Figure 4b), when air was replaced by the LCs, were caused by the increased refractive index of the surrounding media. The decrease in the intensity of extinction efficiency was due to the reduced scattering or reflection of the incident light at the presence of LCs on the nanodisks.

Upon photoirradiation of the Au nanodisks embedded in the LCs with a pump light (a laser beam of 420 nm wavelength and 20 mW power), the LSPR peak made a blueshift ($\Delta\lambda = 30$ nm) with decreased intensity (Figure 5a). Reversible tuning of the LSPR was achieved by

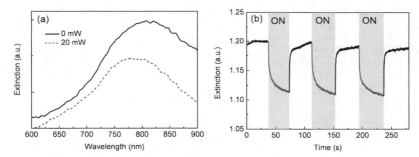

Figure 5. (a) Extinction spectra of Au nanodisks embedded in azobenzene-doped LCs before (solid curve) and after (dashed curve) a pump light was switched on (420 nm wavelength; 20 mW power). (b) Time dependence of extinction at a track wavelength of 725 nm as the pump light was switched on and off alternatively.

switching on and off the pump light alternatively. Figure 5b shows the time dependence of extinction at a track wavelength of 725 nm with the pump light on and off alternatively.

CONCLUSIONS

We have demonstrated a new type of all-optical active plasmonics based on an ordered Au nanodisk array embedded in azobenzene-doped liquid crystals. The function of active plasmonics utilizes the photo-induced phase transition of the doped LCs. This transition modulates the refractive index of the LCs, thereby altering LSPR of the embedded Au nanodisks. The experimental modulation of the LSPR matched well that from DDA calculations. The reversible modulation was repeated by switching on and off the pump light alternatively. This demonstration contributes to the emerging field of plasmonics, an area that will lead to the development of very large scale electronics and photonics integration.

ACKNOWLEDGMENTS

This work was supported by the NSF NIRT grant (ECCS-0609128) and the Penn State Center for Nanoscale Science (MRSEC). The authors also acknowledge use of facilities at the PSU site of NSF NNIN.

REFERENCES

1. E. Ozbay, *Science* **311**, 189 (2006).
2. C. Genet and T. W. Ebbesen, *Nature* **445**, 39 (2007).
3. P. Andrew and W. L. Barnes, *Science* **306**, 1002 (2004).
4. K. J. Chau, M. Johnson, and A. Y. Elezzabi, *Phys. Rev. Lett.* **98**, 133901 (2007).
5. Y. R. Leroux, J. C. Lacroix, K. I. Chane-Ching, C. Fave, N. Felidj, G. Levi, J. Aubard, J. R. Krenn, and A. Hohenau, *J. Am. Chem. Soc.* **127**, 16022 (2005).
6. Z. C.Wang and G. Chumanov, *Adv. Mater.* **15**, 1285 (2003).
7. W. Dickson, G. A. Wurtz, P. R. Evans, R. J. Pollard, and A. V. Zayats, *Nano Lett.* **8**, 281 (2008).
8. Y. B. Zheng, B. K. Juluri, X. L. Mao, T. R. Walker, and T. J. Huang, *J. Appl. Phys.* **103**, 014308 (2008).

Mater. Res. Soc. Symp. Proc. Vol. 1077 © 2008 Materials Research Society

Optimization of Plasmonic Nano-Antennas

Kursat Sendur, Orkun Karabasoglu, Eray Abdurrahman Baran, and Gullu Kiziltas
Sabanci University, Istanbul, 34956, Turkey

ABSTRACT

The interaction of light with plasmonic nano-antennas is investigated. First, an extensive parametric study is performed on the material and geometrical effects on dipole and bow-tie nano-antennas. The transmission efficiency is studied for various parameters including length, thickness, width, and composition of the antenna as well as the wavelength of incident light. The modeling and simulation of these structures is done using 3-D finite element method based full-wave solutions of Maxwell's equations. Next, a modeling-based automated design optimization framework is developed to optimize nano-antennas. The electromagnetic model is integrated with optimization solvers such as gradient-based optimization tools and genetic algorithms.

INTRODUCTION

Nano-optical applications, such as scanning near-field optical microscopy [1] and data storage [2], require intense optical spots beyond the diffraction limit. Nano-antennas [3-4] can obtain very small optical spots, but their ability to obtain optical spots beyond the diffraction limit is not sufficient for practical applications. In addition to a very small optical spot, a nano-antenna should provide high transmission efficiency for practical applications. The transmission efficiency of a nano-antenna determines the data transfer rate of storage devices and scan times of near-field optical microscopes. Therefore, the efficiency of nano-antennas should be optimized for potential utilization in practical applications. Optimization of nano-antennas is crucial for understanding their potential and limitations for emerging plasmonic applications.

The interaction of antennas with electromagnetic waves has been thoroughly investigated at microwave frequencies. Scaling and optimization rules do not apply at optical frequencies [4]. At visible and infrared frequencies the underlying physics of the interaction of light with metallic nano-antennas is complicated due to the behavior of metals as strongly coupled plasmas [5-8]. Experimental studies have shown light localization using both dipole [9] and bow-tie [10] nano-antennas. A brute-force optimization study of these structures is not practical due to large number of parameters. There is a need for a systematic optimization of these structures.

In this study, we develop a modeling-based automated design optimization framework to optimize nano-antennas. The modeling and simulation are done using 3-D finite element method (FEM), which is integrated with optimization solvers such as genetic algorithms and gradient based optimization tools. First, an extensive parametric study is performed on the material and geometrical effects. Then the proposed design framework is used to optimize nano-antennas.

DIPOLE AND BOW-TIE PLASMONIC NANO-ANTENNAS

To couple incident electromagnetic energy with small scale electronic devices, antennas have been utilized. The antennas achieve this coupling by localizing the incident radiation to smaller dimensions than the wavelength. This coupling mechanism has been well understood for radio frequency and microwave applications, and is also applicable to nano-antennas operating at optical frequencies. At optical frequencies nano-scale metallic antennas can be utilized to couple incident optical beams to length scales much smaller than the diffraction limit.

An antenna is composed of metallic parts. For example, the dipole antenna shown in Fig. 1 (a) is composed of two metallic rods separated by a distance, G. Similarly, a bow-tie antenna shown in Fig. 1 (b) is composed of two triangular metallic pieces, which are also separated by a distance, G. The incident electromagnetic wave creates induced currents on the antenna surface. These induced currents are the source of charge accumulation at the ends of the antenna. If the incident polarization is along the long-axis of the antenna, then the charges of opposite sign are created across the gap separating the metallic parts. The oscillation of the charge is the main source of a localized near-field electromagnetic radiation. This localized radiation is composed of propagating and evanescent components. If a structure is brought in the near-field of this gap, then the radiating fields due to this charge oscillation interact with the structure, which leads to interesting applications, including near-field optical microscopy [1] and data storage [2].

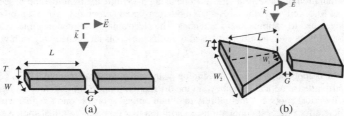

Figure 1. A schematic illustration of a (a) dipole and (b) bow-tie antenna, and their dimensions. The antennas are illuminated with incident electromagnetic radiation shown with $\vec{E}$.

At visible and infrared frequencies, however, the underlying physics of the interaction of light with metallic nano-antennas is complicated due to the behavior of metals as strongly coupled plasmas [5-8]. If the shape and composition of the metallic structures and the wavelength of the incident radiation is appropriately chosen, then it is possible to excite surface plasmon resonances over these metallic particles [11]. Size and shape-dependent surface plasmon resonances of nano-particles can be excited if the frequency of the incident radiation matches the frequency of the oscillation frequency of the free-electron gas of metals.

Surface plasmon resonances are associated with high electric field enhancement. This field enhancement is of particular interest for applications that require high transmission efficiencies in addition to small optical spots. One way to further enhance the electromagnetic near-field radiation is to utilize sharp tips. Due to the lightning rod effect around the sharp tips, the electric field enhancement, and therefore the transmission efficiency of the nano-antenna can be further improved. Therefore, the sharp tips of the bow-tie antenna should provide better field improvement compared to the dipole antenna. However, for a more realistic simulation the bow-tie antennas are modeled as shown in Fig. 1, where the sharp tips are truncated at a width of W_1.

We performed numerical simulations using the FEM to understand the effect of various parameters on the near-field radiation of nano-antennas. Radiation boundary conditions are used in FEM simulations. Tetrahedral elements are used to discretize the computational domain. On the tetrahedral elements, edge basis functions, and second-order interpolation functions are used to expand the functions. Adaptive mesh refinement is used to improve the coarse solution regions with high field intensities and large field gradients. The optical properties of metals used in this study are retrieved from the literature [12]. In Fig. 2 (a), electric field distribution for a dipole antenna is shown on the x-z cut-plane plane, which passes through the center of the antenna. The incident field is polarized in the x-direction, and is propagating in the negative z-direction. The wavelength of the incident light is 850 nm. The result shows a confined electric field close to the gap region of the antenna. Also a large electric field enhancement is observed. For this simulation, the sizes of the antenna are L= 110, T= 20, W= 20, and G= 20 nm. The magnitude of the incident field is 1 V/m for this calculation. The result suggests that the intensity enhancement at the center of the antenna is over 1000.

In Fig. 2 (b) we next plot the intensity, $|E|^2$ as a function of the length of the antenna. To achieve this plot, we simulated the near-field radiation from the antenna for different antenna lengths, and recorded the electric field values as shown in Fig. 2. (b). The results suggest that the optimum length is around 90 nm when the incident wavelength is 850 nm. Note that the length is for one of the metallic rods. Including the gap and the other rod, the entire length is 200 nm.

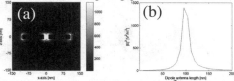

Figure 2. (a) The intensity distribution, $|E(x, y=0, z)|^2$ on the x-z plane for a dipole antenna at a wavelength of 850 nm for L= 110, T= 20, W= 20, and G= 20 nm, and (b) The intensity at the center of the gap $|E(x=0, y=0, z=0)|^2$ is plotted for various L at λ = 850 nm.

A similar calculation is repeated for a bow-tie antenna in Fig. 3 (a). Similar to the previous set of calculations, the electric field distribution for a bow-tie antenna is shown on the x-z cut-plane, which passes through the center of the antenna. The wavelength of the incident field is 850 nm. The dimensions of the bow-tie antenna are L= 100, T= 20, W_1= 20, W_2= 220, and G= 20 nm. The field enhancement is slightly larger than that of the dipole antenna. The magnitude of the incident field is 1 V/m for this calculation. The intensity, $|E|^2$, enhancement at the center of the antenna is larger than 1200 for this simulation. In Fig. 3 (b) we next plot the electric field enhancement as a function of the length of the antenna. The results suggest that the optimum length is 120 nm when the incident wavelength is 850. Note that the length is for one of the triangular structures. Including the gap and the other triangle, the entire length is 260 nm.

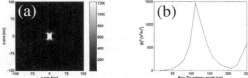

Figure 3. (a) The intensity distribution, $|E(x, y=0, z)|^2$ on the x-z plane for a bow-tie antenna at a wavelength of 850 nm for L= 100, T= 20 , W_1= 20 , W_2= 220, and G= 20 nm, and (b) The intensity at the center of the gap $|E(x=0, y=0, z=0)|^2$ is plotted for various L, at a λ = 850 nm.

The electric field is plotted at the center of the antenna as a function of the incident wavelength and antenna length. The incident wavelength is varied from 400 nm to 2000 nm by intervals of 50 nm. At each wavelength, we performed simulations by changing the antenna length. The intensity at the center of the gap, $|E(x=0, y=0, z=0)|^2$, is calculated for each wavelength and antenna length. By recording the intensity over the rectangular grid shown in Fig. 4, we formed the surface graphs. Rather than using a constant incident field, we used a constant power value of 1 mW. The power calculations are based on a focused beam model [13]. In Fig 4 (a), the intensity is plotted as a function of wavelength and antenna length for a gold antenna with other parameters set as T= 10, W= 10, and G= 20 nm. The optimum wavelength shifts toward longer wavelengths as the antenna length is increased. There is a sharp drop in the intensity enhancement due to surface plasmon damping. Fig 4 (b) shows the results for T= 20, W= 20, and G= 20 nm. Increasing the antenna cross section shifts the optimum wavelengths to smaller values. A similar trend is observed in Fig 4 (c) with T= 50, W= 50, and G= 20 nm. In Fig. 4 it should be noted that the $|E(x=0, y=0, z=0)|^2$ values do not correspond to the intensity enhancement. The magnitude of the incident electric field in not 1 V/m, rather the incident power is selected as 1 mW. Intensity enhancement is still in the order of 1000 as shown in Figs. 2 and 3.

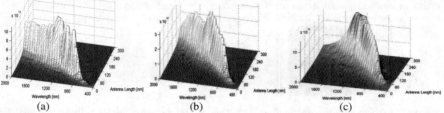

(a)   (b)   (c)

Figure 4. The intensity as a function of wavelength and antenna length for: (a) T= 10, W= 10, and G= 20 nm, (b) T= 20, W= 20, and G= 20 nm, and (c) T= 50, W= 50, and G= 20 nm.

The important parameters in Fig. 4 are the optimum antenna length and the corresponding intensity value at various wavelengths. Figure 5 (a) and (b) illustrate the optimum length and corresponding intensity values extracted from surface plots for gold dipole antennas with various cross sections. In Fig. 5 (a) and (b) the optimum length of the antenna and the corresponding intensity values are plotted as a function of the wavelength. The results in Fig. 5 (a) suggest that the optimum antenna length is longer for thicker antennas. The optimum length and intensity are plotted for a silver dipole antenna as a function of the wavelength in Fig. 6.

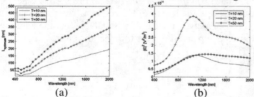

(a)   (b)

Figure 5. (a) Optimum gold dipole antenna length as a function of wavelength, (b) Optimum intensity values $|E(x, y=0, z)|^2$ for optimum antenna lengths as a function of wavelength.

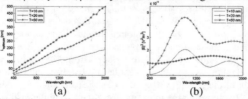

(a)   (b)

Figure 6. (a) Optimum silver dipole antenna lengths as a function of wavelength, (b) Optimum intensity values $|E(x, y=0, z)|^2$ for optimum antenna lengths as a function of wavelength.

In Fig. 7, the optimum length of a gold bow-tie antenna and the corresponding intensity value at various wavelengths is plotted. The bow-tie antenna parameters are selected as T= 20, W_1= 20, and G= 20 nm. The flare angle is selected as 45°, therefore, the width W_2 corresponds to 20+2*L as the length of the antenna is changed. The results show a trend similar to dipole antennas. The optimum antenna length is, however, longer than that of a dipole antenna. At shorter wavelengths, some unexpected spikes are observed as shown in Fig. 7. In Figs. 2 (b) and 3 (b), if we continue to increase the length of the antenna, we observe further maxima and minima. These spikes correspond to the second optimum length of the antennas. The second and higher optima should provide lower intensity values. However, due to numerical errors, the second maximum gives slightly higher intensity compared to the first maximum at these spikes.

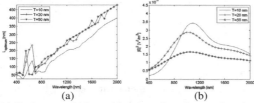

(a)   (b)

Figure 7. (a) Optimum gold bow-tie antenna lengths as a function of wavelength, (b) Optimum intensity values $|E(x, y=0, z)|^2$ for optimum antenna length as a function of wavelength.

OPTIMIZATION FRAMEWORK

The surface plasmon resonances of nano-antennas depend on parameters related to the shape and composition of the nano-antenna as shown in Fig. 1. Complete understanding of surface plasmon resonances of nano-optical systems requires a complete and detailed understanding of possibly many more design parameters, geometries, and material properties. The large number of parameters involved in studying functional plasmonic devices with a brute force numerical parameter simulation is not feasible. To design novel nano-optical transducers a modeling based automated design optimization framework is necessary.

The design framework is formed by integrating a commercial electromagnetic analysis tool Ansoft HFSS with MATLAB's optimization toolbox. Specifically, two different optimization tools are integrated on a MATLAB based scripting interface to iteratively search for optimum geometric parameters of a dipole and bowtie antenna: sequential quadratic programming (SQP) and genetic algorithm (GA). The optimization model consists of maximizing the field intensity $|E(x=0, y=0, z=0)|^2$ subject to bound constraints of [20, 450] and [400 2000] for geometric length and wavelength, respectively. Convergence is achieved in about less than 20 iterations and 10 generations for the SQP and GA framework, respectively. Optimization parameters in the GA setting include 10 individuals, Gaussian Mutation and Roulette Wheel Selection. Optimal lengths for dipole antennas are obtained via SQP, plotted with respect to wavelength, and compared to results obtained via the brute-force simulation in Figure 8. There is an overall agreement except for optimal lengths at wavelengths close to bound constraints. The discrepancies are attributed to two main reasons: Inaccurate brute-force predictions of maximum field intensity of finite sampled frequency points and as expected with gradient based optimization tools, results show that SQP's performance in locating the optimum solution depends on the chosen initial design with especially when the intensity is a multi-modal function. The GA based optimization framework seems to overcome this issue in the expense of

131

computational time. Optimal results for the bow-tie antenna length converged to 140 nm at 900 nm, and to a dipole length of 286 nm at 1764 nm for a 2 variable optimization study via the GA framework while the SQP was unable to converge for the latter. Initial results seem to be promising in providing the capability of exploring nano-structures with several design parameters. The electric field performance is likely to result in more complicated response functions. Future work includes expanding the framework to hybridize both optimization tools in combining the advantages of global and local optimization tools and to expand the framework to multi-objective design optimization problems.

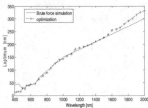

Figure 8. Comparison of the optimization result for a dipole antenna using the SQP method and brute-force simulations.

CONCLUSIONS
In this study, interaction of light with plasmonic nano-antennas was investigated. An extensive study is performed to investigate the effect of the geometric and material properties of nano-antennas on the transmission efficiency. A modeling based automated design optimization framework was also developed. The results of the optimization framework were compared with those of the brute-force simulations.

REFERENCES
1. A. Hartschuh, E. J. Sanchez, X. S. Xie, and L. Novotny, *Phys. Rev. Lett.*, **90**, 095503 (2003).
2. K. Sendur, W. Challener, and C. Peng, *J. Appl. Phys.*, **96**, 2743-2752 (2004).
3. R. D. Grober, R. J. Schoelkopf, and D. E. Prober, *Appl. Phys. Lett.*, **70**, 1354 (1997).
4. L. Novotny, *Phys. Rev. Lett.*, **98**, 266802 (2007).
5. K. Sendur and W. A. Challener, *J. Microsc.*, **210,** 279-283 (2003).
6. H. Raether, *Surface Plasmons on Smooth and Rough Surfaces and on Gratings* (Springer Verlag, 1988).
7. J. J. Burke, G. I. Stegeman, and T. Tamir, *Phys. Rev. B*, **33**, 5186 (1986).
8. H. Raether, *Physics of thin films*, vol. 9, pp. 145-261 (Academic, New York, NY, 1977).
9. P. Muhlschlegel, H.-J. Eisler, O. J. F. Martin, B. Hecht, and D. W. Pohl, *Science*, **308**, 1607-1609 (2005).
10. F. Jackel, A. A. Kinkhabwala, and W. E. Moerner, *Chem. Phys. Lett.*, **446**, 339-343 (2003).
11. K. L. Kelly, E. Coronado, L. L. Zhao, and G. C. Schatz, *J. Phys. Chem. B*, **107**, 668-677 (2003).
12. E. D. Palik, *Handbook of optical constants of solids* (Academic Press, San Diego, CA, 1998).
13. B. Richards and E. Wolf, *Proc. Roy. Soc. London Ser. A*, **253**, 358-379, (1959).

Mater. Res. Soc. Symp. Proc. Vol. 1077 © 2008 Materials Research Society 1077-L04-01

Fabrication of Nano-Structured Gold Arrays by Guided Self-Assembly for Plasmonics

Xiaoli V. Li[1], Clelia A. Milhano[2], Robin M. Cole[3], Phil N. Bartlett[2], Jeremy J. Baumberg[3], and Cornelis H. de Groot[1]

[1]School of Electronics and Computer Science, University of Southampton, Southampton, SO17 1BJ, United Kingdom

[2]School of Chemistry, University of Southampton, Southampton, SO17 1BJ, United Kingdom

[3]NanoPhotonics Centre, University of Cambridge, Cambridge, CB3 0HE, United Kingdom

ABSTRACT

Gold inverse spherical nanoscale voids have been fabricated in linear arrays for directional plasmon measurements in the visible spectral range. We show that by KOH anisotropic etching in Si, we are able to make V-grooves in which latex spheres of the order of 500 nm self-assemble with largely defect-free cubic symmetry. Both single layer and multilayer assembly in a face-centered close-packed (FCC) lattice can be achieved by varying the width of the trenches. This template is subsequently used for electrodeposition of gold to create the inverse spherical nanovoids.

INTRODUCTION

Monodispersed spherical colloids can self-assemble into one-dimensional to three-dimensional lattices [1–3] by various types of driving forces, such as gravity [4], convection [5], spin-coating [6] or electrostatics [7]. To achieve long-range well-ordered colloidal lattices, a promising approach is guided self-assembly, in which colloidal spheres self-assemble onto a patterned substrate [8, 9]. For colloidal spheres, the main packing force is the interactions between spheres, which is often non-directional, in a sense, very similar to ionic bonding and metallic bonding. On a flat substrate, they self-assemble into face-centered or hexagonal close packed monolayer or multi-layers because this type of packing has the maximum density [10]. However, on a patterned substrate, the physical constraint, which is the volume interactions between spheres and the walls of the templates, is the main packing force. Since it is directional, the packing stucture of colloidal crystals on patterned substrates is based primarily on physical constraints, as shown in Ref. [11-15].

These high quality three dimensional colloidal sphere arrays have attracted attention in photonic applications, such as waveguide structures, optical filters and switches because of their potential as template for the fabrication of photonic crystals. In particular, metal inverse spherical nanoscale voids are useful because they possess plasmonic modes significantly different from those of metal nanoparticles. Submicron spheres are demanded for visible light plasmonic study. The dimensions of the spheres that were self-assembled in Ref. [11-15] were too large to be suitable for visible light study. In previous work, hexagonally close-packed spherical nanoscale voids have been fabricated by ourselves through self-assembled latex sphere templates by metal electrodeposition. The inverse spherical nanovoids support both propagating and localized plasmon modes [16, 17]. However, it was not possible to control the direction of plasmon propagation using this geometry.

In this paper, we present a unique fabrication of gold inverse spherical naonovoids in strip arrays in order to avoid the six-fold symmetry of the dispersion and to guide plasmon propagation in a specific direction. The main fabrication challenge of this work is the combination of self-assembly, metal electrodeposition onto semiconductor and semiconductor fabrication techniques. We show the gold strip arrays are of high enough quality to allow investigation of plasmonic signals.

EXPERIMENT

The templates were made of monodisperse polystyrene latex spheres (Duke Scientific Corporation) supplied as a 1 wt.% solution in water (manufacturer's certified mean diameter of 499 nm±5 nm, coefficient of variation in diameter 1.3%). Before use, the suspensions were homogenized by successive, gentle inversions for a couple of minutes followed by a sonication for 15 s. All solvents and chemicals were of reagent quality and were used without further purification.

The process flow for the samples is outlined in Figure 1. The pre-patterned silicon substrates were prepared from n-type (100)-polished silicon substrate wafers with 0.01~0.02 $\Omega \cdot$cm resistivity. A layer of silicon dioxide with 250 nm thickness was thermally grown and a photoresist layer was spun on top. Whereafter, a conventional photolithography pattern was transferred onto the wafer surface by a Nikon NSR-2005/i9C step and repeat system. Arrays with width varying from 500nm to 1700nm, with step size 100nm, were fabricated with the separation between the strips either being 500nm or 1000nm. The length of each strip is 400 μm in all cases. The oxide was dry etched until the silicon and then the photoresist layer was removed by a fuming nitric acid clean. To obtain V-shaped groove trenches, silicon KOH etching was carried out. A 20:1 BHF dip followed, leaving the Si surface H - terminated and hence, hydrophobic. Latex spheres of 500 nm diameter were self-assembled on the substrates by slow evaporation of a colloidal water suspension containing 1wt.% of latex spheres. Electrodeposition of gold was performed at ambient temperature using a conventional three-electrode configuration controlled by an Autolab PGSTAT12. Just before electrodeposition, another 20:1 BHF dip (6 seconds) was necessary to remove the native oxide and leave the Si surface H-terminated. The sample was the working electrode with a platinum gauze counter electrode and a saturated calomel reference electrode (SCE). A pulse of -1.1 V vs. SCE was applied for 0.2 second just before the deposition stage. Gold inverse spherical nanovoids were deposited under potentiostatic conditions at -0.7 V vs. SCE. The insulating SiO_2 forced electrodeposition to take place selectively on the Si patterns. After electrodeposition, the sample was placed into dimethylformamide (DMF) solvent and washed in an ultrasonic tank for 2 hours in order to dissolve the latex sphere template. The morphology and nanostructures of both the colloidal templates and gold nanovoids were characterized by using Scanning Electron Microscopy (LEO 1455VP SEM).

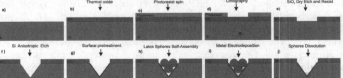

Figure 1. Process flow for guided self-assembly of inverse sphere metal arrays guided by V-shaped trenches.

DISCUSSION

Guided Self-assembly of colloidal spheres

When the samples appear opalescent, as expected, with colors from green to red, depending on the angle of observation with the samples illuminated from above, it indicates the template order is robust with good adhesion to the patterned substrates. In most of the cases, trapping of the spheres into the morphological patterns of the substrate resulted in selective spheres self-assembly on Si.

As shown in fig. 2, colloidal spheres with diameter of 500 nm are observed to form in the V-shaped groove trenches with 600 nm in width as a single-file contiguous arrangement. These linear constructions are located in the apex of the V-shaped groove. All spheres are bound to the neighbours and the assembly is close to perfect for an area of 0.5 mm by 0.2 mm. The spheres share two contact points with the trench wall providing the vertical stability, while the neighouring spheres provide the lateral stability. If the trench width is larger, the assembly is of a more complicated nature until the width of the trench reaches 1200 nm. As shown in fig. 3, a well-ordered cubic packing is visible in the trenches. Not visible in the SEM, is the sphere below the top layer which causes the cubic ordering. The schematics show the cross section of this structure. Spheres in the bottom layer have two pairs of interactions: wall to sphere and sphere to sphere while the spheres in the second layer have three types of interactions: a pair of sphere to sphere interactions from the adjacent bottom spheres, a pair of sphere to sphere interactions from the longitudinal adjacent spheres in same layer and one force from the sidewall of Si<111>. It is noteworthy that <100> layer planes are well ordered and organized parallel to the <100> face of the single crystal Si<100> wafer. As shown in fig. 4, for colloidal spheres with 500 nm diameter, a third layer and a fourth layer are visible in the trenches respectively. In fig. 4(a), the exposed double chain layer in the right trench reveals the invisible terrace structures under top layer. The commensurate width of the trench was n times of 600 nm, with n being an integer. Note that the number of spheres observed in the top <100> layer corresponds to the number of sphere

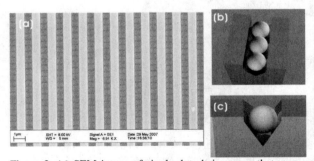

Figure 2. (a) SEM image of single dot chain arrays that were assembled by 500 nm colloidal spheres into trenches with 600 nm in width. (b,c) Top view and cross section view of schematic structures.

layer planes of the colloidal crystal within the V-shaped grooves. So that once the widths of the trenches were commensurate with the diameter of spheres, the <100> colloidal crystal would grow up until the trenches were filled completely. Therefore, due to the 70.6° angular geometry of the V-shaped grooves, in the multilayer modes, colloidal spheres were collected and self-assembled as hexagonally close-packing lattices on the trench wall to achieve the highest packing density and confined by the grooves to form the colloidal crystal parallel chains.

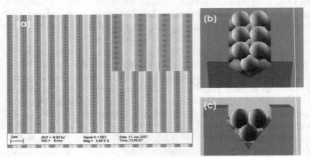

Figure 3. (a) SEM image of double dot chain arrays that were assembled by 500 nm colloidal spheres into trenches with 1200 nm in width. (b,c) Top view and cross section view of schematic structures.

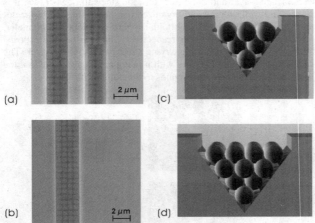

Figure 4. (a,b) SEM image of 3D <100> oriented lattices that were assembled by 500 nm colloidal spheres into trenches with 1800 nm and 2400nm in width, respectively. (b,c) Cross section of schematic structures.

Au Metallization

After self-assembly, the 3D lattices were used as templates to fabricate gold inverse spherical strip arrays on Si. A pulse of -1.1 V vs. SCE for 0.2 s is used to form a layer of uniform Au nucleation. With semiconductor Si as the electrodeposition interface, instantaneous uniform nucleation is the key step to obtain a continuous gold film [18, 19]. The selected potential of -0.7 V vs. SCE controls the electrodeposition process to carry out slowly growth without any further nucleation. As shown in fig. 6, gold inverse spherical nanovoids are grown along the single dot chain and double dot chain templates that were assembled from 500 nm diameter colloidal spheres in arrays of trenches with 700 and 1400 nm in width, respectively. Gold electrodeposition occurs only on the Si surface but not on the insulating SiO_2, resulting in confinement of the metal arrays exclusively to the Si patterns. Gold begins to grow from both walls of V-shaped trenches, where the orientation of silicon is <111>. Due to the 70.6° angular geometry, the Au film from both sides of walls joints together from the apex of the trenches. Because of the superposition of gold growth surface, the voids are gradually filled in by gold metal starting from the bottoms. Compared with the nanovoids grown on flat substrate, the extraordinary structure of these grown into V-shaped groove trenches are the flowering distributions of gold density. The thickness of the gold layer is controlled by estimation of the applied charge during electrodeposition. Corresponding to different thickness of Au deposition, the void shapes range from shallow dishes, to triangular islands, to truncated spherical cavities.

CONCLUSIONS

We have demonstrated a method to fabricate gold inverse spherical nanovoids arrays. In this method photolithography is used to pre-pattern a silicon surface, which is then selectively etched in KOH to create series of V-shape groove trenches. The geometric confinement of the trenches

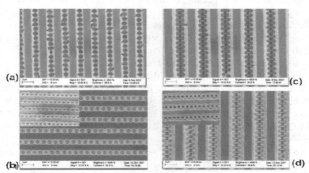

Figure 5. SEM images of single and double dot chains templated gold arrays. These gold nanovoides are grown along the 3D template that was assembled from 500 nm colloidal spheres in arrays of trenches with 700nm (a,b) and 1400 nm (c,d) in width, respectively. (a,c) With sphere templates. (b,d) After spheres dissolved. Inset was the view of 45± angle tilted.

together with electrostatic interactions guides the assembly of the spheres on this patterned silicon surface. Zero-defect arrays were found to assemble along the entire length of some trenches, leading to state of the art defect-free areas of 0.5×0.2 mm^2 for 500nm diameter spheres, allowing characterisation with visible light. The array formation was found to be highly dependent on the width of the trenches. Various array formations, including single dot chains, parallel chains (cubic packing) and helical chains, were achieved by modifying the trench width. These long-range well-ordered sphere strip arrays are used as template for metallization to form gold nanovoids strip arrays. By varying the film thickness, void shapes ranging from shallow dishes to encapsulated voids were fabricated. The gold strip arrays are of high quality to satisfy the detection of plasmonic signals. Further understanding of the different plasmon modes will aid research into slow-light and active waveguides and potentially allow on-chip optical communication.

REFERENCES

1. C. Bae, H. Shin, and J. Moon, Chemistry of Materials, **19**, 1531 (2007).
2. J. Ye, R. Zentel, S. Arpiainen, J. Ahopelto, F. Jonsson, S.G. Romanov, and C.M. SotomayorTorres, Langmuir, **22**, 7378 (2006).
3. K. Nagayama, Colloids Surfaces A, **109**, 363 (1996).
4. L.N. Donselaar, A.P. Philipse, and J. Suurmond, Langmuir, 13(23):6018–6025, 1997.
5. N.D. Denkov, O.D. Velev, P.A Kralchevsky, I.B. Ivanov, H. Yoshimura, and K. Nagayama, Langmuir, 8:3183–3190, 1992.
6. J. C. Hulteen and R.P. van Duyne, J. Vac. Sci. Technol. A, 13:1553–1558, 1995.
7. A. M. Kalsin, M. Fialkowski, M. Paszewski, S. K. Smoukov, K. J. M. Bishop, and B. A. Grzybowski, Science, 312(5772):420–424, 2006.
8. Alfons van Blaaderen, Rene Ruel, and Pierre Wiltzius, Nature, 386:321–324, 1997.
9. M. E. Kiziroglou, X. Li, D. C. Gonzalez, C. H. de Groot, A. A. Zhukov, P. A. J. de Groot, and P. N. Bartlett, J. Appl. Phys., 100:113720–113725, 2006.
10. J. H. Conway and N. J. A. Sloane. Sphere packings, lattices and groups. Springer-Verlag, 1988.
11. G. A. Ozin and S. M. Yang, Advanced Functional Materials, 11:95–104, 2001.
12. San Ming Yang and Geoffrey A. Ozin, Chem. Commun., pages 2507–2508, 2000.
13. Yin Yadong, Lu Yu, Gates Byron, and Y. Xia, J. Am. Chem. Soc., **123**, 8718 (2001).
14. Yin Yadong, Li Zhi-Yuan, and Younan Xia, Langmuir, 19:622 – 631, 2003.
15. Younan Xia ; Yadong Yin; Yu Lu, Advanced Functional Materials, **13**, 907, 2003.
16. T. A. Kelf, Y. Sugawara, R. M. Cole, J. J. Baumberg, M. E. Abdelsalam, S. Cintra, S. Mahajan, A. E. Russell, and P. N. Bartlett, Physical Review B (Condensed Matter and Materials Physics), 74(24):245415, 2006.
17. R.M. Cole, J.J. Baumberg, F.J. GarciadeAbajo, S. Mahajan, M. Abdelsalam, and P.N. Bartlett, Nano Letters, 7(7):2094–2100, 2007.
18. M. E. Kiziroglou, A. A. Zhukov, M. Abdelsalam, X. Li, P. A. J. de Groot, P. N. Bartlett, and C. H. de Groot, IEEE Transactions on Magnetics, 41:2639–2641, 2005.
19. M. E. Kiziroglou, A. A. Zhukov, X. Li, D. C. Gonzalez, P. N. de Groot, P. A. J.and Bartlett, and C. H. de Groot, Solid State Communications, 140:508–513, 2006.

Mater. Res. Soc. Symp. Proc. Vol. 1077 © 2008 Materials Research Society   1077-L06-04

Silicon-Compatible Ultra-Long-Range Surface Plasmon Modes

Ali Sabbah[1], C. G. Durfee[1], R. T. Collins[1], T. E. Furtak[1], R. E. Hollingsworth[2], and P. D. Flammer[1]

[1]Colorado School of Mines, Golden, CO, 80401
[2]ITN Energy Systems, Littleton, CO, 80127

ABSTRACT

It has long been known that the range of surface plasmons can be extended by sandwiching a thin metal film between dielectrics of equal refractive index. We have modeled the effect of breaking the symmetry of this structure by adding next to the metal film a thin dielectric layer with lower refractive index than the outer dielectric layers. With careful control of the low index layer thickness and dielectric constant, a bound surface plasmon mode with an even greater propagation length can be created for a metal film of finite thickness. We have experimentally confirmed the existence of these modes using an attenuated total reflection measurement to study surface plasmon modes in a Pyrex/Ag/MgF$_2$/oil structure where the oil is index matched to the Pyrex. Shifts in the mode observation angle and mode half width with the MgF$_2$ thickness agree with model predictions. This ultra-long range surface plasmon is of particular interest because a waveguide structure that supports it can have a silicon compatible, metal-oxide-semiconductor (MOS) configuration.

INTRODUCTION

Electromagnetic waves bound to metal interfaces, or surface plasmons (SPs), have recently attracted considerable interest for their potential to enable sub-wavelength integrated optical devices. SP-based waveguides and modulators would be particularly useful if they could be built and modulated using well developed, MOS compatible Si microelectronics processing allowing a truly integrated optoelectronic technology to be developed. Since silicon lacks an electro-optic effect, efforts to build electrically controlled modulators on silicon (both conventional and plasmonic) are exploring electrically induced free carrier modulation of the dielectric constant in the Si adjacent to a waveguide [1]. This has two major impacts on the design of plasmonic modulators. First, since the dielectric constant modulation is relatively small, the waveguide structure must have very low loss (a long propagation length) to allow sufficient interaction distance for the free carrier modulation to be effective. Second, silicon structures consistent with free carrier modulation, such as metal-oxide-silicon (MOS) capacitors where the carrier concentration under the oxide can be modulated, introduce an asymmetry into the plasmonic waveguide structure. Such asymmetries have previously been reported to complicate SP waveguide performance [2, 3].

Since long propagation length requires limiting the damping of SP modes by the metal layer, it is important to develop devices in which the electric fields associated with the SP are localized outside of the metal. Excitations of this type were predicted and demonstrated (although not in silicon) nearly 25 years ago and are commonly referred to as long range surface plasmon (LRSP) waveguide modes. The LRSP excitation involves an interaction of surface plasmons propagating on opposite surfaces of a thin metal film surrounded by dielectric layers with the same dielectric constant. The modes interact with their E-fields nominally parallel or anti-parallel to create two bound modes. The anti-parallel combination leads to an energy density node within the metal, resulting in lower ohmic loss. Ideally, as the thickness of the film goes to zero, the propagation length becomes infinite. LRSPs, however, have several limitations for silicon based waveguides and modulators. First, LRSP losses increase substantially as the refractive index of the surrounding medium increases because the electric field is pushed back into the metal. If the surrounding material is silicon, useful propagation lengths require unrealistically thin metal layers. Second, to produce a charge accumulation or inversion layer next to the metal surface without substantial power loss, a gate oxide will likely separate the metal from the silicon. This creates an asymmetry in the structure. It has previously been argued that such asymmetries push guided SP modes past cutoff [2].

MODELING

We have used a transfer matrix approach [4] to explore SP modes in asymmetric structures. We find that the asymmetry introduced by a low-index oxide adjacent to the metal layer can, in fact, lead to a bound SP mode with ideal propagation length greater than 1mm. We term this an ultra-long range surface plasmon or ULRSP to distinguish it from the conventional LRSP. The thickness and refractive index of the low-index layer offers a measure of control over the SP propagation length, allowing the use of less-than-ideal metals (such as the MOS-compatible metals Al or Cu). We have explored a variety of structures to illustrate the influence of the oxide layer on waveguide design. The behavior of the LRSP mode and two ULRSP configurations, for example, is illustrated in Fig. 1, which shows the real (n'_{sp}) and imaginary (n''_{sp}) parts of the effective mode index as a function of the thickness (h_m) of the metal layer. The imaginary part of the index is inversely related to the loss in the mode and hence determines propagation length. The red curves correspond to a conventional LRSP structure composed of a silver layer embedded in silicon with $\lambda_0 = 1.55$ μm. For large h_m there is a single value for the real and imaginary indices. As h_m decreases, the mode splits into high-loss (a_b) and low-loss (s_b) branches, where the b subscript indicates these are bound modes. The low loss s_b branch corresponds to the LRSP mode. For the s_b branch the size of n''_{sp} goes to zero (and its range goes to infinity) only in the limit of $h_m = 0$.

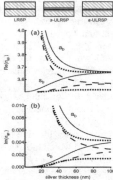

Figure 1. (a) Real part n'_{sp} and (b) Imaginary part n''_{sp} of the effective index at $\lambda_0 =1.55$μm of surface plasmon modes versus metal (Ag) layer thickness (h_m) for different structures. Line curves: (Si)/(Ag)/(Si), LRSP; : Dashed curves: (Si)/(7.5nm TiO$_2$)/(Ag)/(7.5nm TiO$_2$)/(Si), s-ULRSP. Dotted curves: (Si)/(15nm TiO$_2$)/(Ag)/(Si), a-ULRSP. The lower (upper) curves correspond to the H-symmetric, s_b (H-antisymmetric, a_b) solutions.

The figure illustrates two other configurations where a lower index oxide layer has been introduced on one (a-ULRSP, dotted curves) or both (s-ULRSP, dashed curves) sides of the metal film. The metal and oxide layers are surrounded by Si. In these calculations, the index of the oxide is $n_{ox} = 2.5$ (corresponding to TiO_2 at 1.55 µm). Both configurations would be consistent with electrically forming a charge accumulation or inversion layer in the silicon adjacent to an oxide layer, i.e. an MOS design. Focusing on the a-ULRSP configuration, for large h_m, the modes are uncoupled and correspond to isolated SPs at the two different interfaces. As h_m decreases the modes begin to couple. Unlike the LRSP case, for this configuration it is possible to reach zero losses in the s_b mode at a finite value of h_m. Here we describe the mode as H-symmetric in analogy to the LRSP nomenclature. In fact, the mode is not symmetric. This ULRSP behavior can also be obtained with symmetric placement of two low-index layers (dashed curve) [5].

Figure 2 shows the H-field profile for a conventional LRSP mode and for the low loss a-ULRSP configuration. As the oxide thickness is increased, the field is drawn toward the oxide side of the structure and the mode becomes less confined while the propagation length increases. The oxide thickness and refractive index are both involved in giving ultra-long range propagation. For fixed Ag layer thickness, thicker oxide generally increases the mode size and propagation length until a cutoff condition is reach where the mode is no longer bound. For fixed oxide thickness, a higher index oxide decreases both mode size and propagation length. This means that mode size and propagation length can be tuned to a specific application independent of metal film thickness by controlling oxide properties. This is quite useful in the optimization of waveguide based modulators. It is also noteworthy the field in the a-ULRSP structure is enhanced on the oxide side of the metal which is the region where free carrier induced index modulation will occur, while the field is reduced on the opposite side which is likely to be lower quality deposited silicon with higher loss. (Our modeling used perfect crystalline silicon on both sides so the role of the oxide could be isolated.)

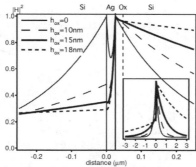

Figure 2. H-field intensity profiles for constant 25nm Ag thickness and several oxide thicknesses as indicated. Solid vertical lines show Ag boundaries; dashed vertical line shows a representative oxide/Si boundaries. The propagation lengths for 0nm, 10nm, 15nm, and 18nm oxide thicknesses are 0.26mm, 0.39mm, 0.82mm, and 2.75mm, respectively. Inset shows same fields over a larger horizontal scale.

EXPERIMENT

To demonstrate the existence and confirm the properties of ULRSP modes we have studied a visible wavelength analog of the a-ULRSP structure. SP modes were detected using reflectance measurements in ATR configuration with visible light from a HeNe laser (633nm) as shown in Fig. 3. The structure used in this experiment was oil/Ag/MgF$_2$/Pyrex where the oil is index matched to Pyrex creating an analog to Si/Ag/oxide/Si structures discussed above. The structure was fabricated beginning with an optically polished Pyrex glass substrate (n=1.47). A thin layer of MgF$_2$ (n=1.388, t=94nm) was deposited on the substrate to play the role of the oxide layer in the MOS structure. A thin silver film, 61.5nm, was then evaporated on top of the MgF$_2$. A control sample was prepared without the MgF$_2$ layer and with the Ag evaporated directly on the Pyrex. Optical constants of the silver film were measured with an ellipsometer to be ε_{Ag} = (-17.308, 0.188) at 633nm. An index matching fluid with nearly the same index as the Pyrex was then used to optically couple the structure to a high index glass prism (SF10, n=1.7231) as illustrated in Fig. 3. The coupling dielectric fluid thickness was tuned by applying pressure between the substrate and prism.

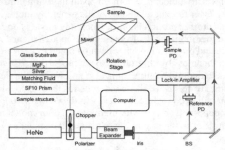

Figure 3. Experimental set up for the ATR measurements. The HeNe laser beam is being chopped for synchronized detection and polarized before entering the prism to excite SP on the sample on the prism side. The reflected beam from the sample and the reference beam are detected with silicon photodiodes and recorded on a computer through a lock-in amplifier. The sample structure (Silver/MgF2/Gass) is an asymmetric structure with analogy to metal-oxide-semiconductor (MOS).

RESULTS

The prism coupling allows surface plasmons to be excited at a specific incidence angle when the component of the wave vector of the incidence light parallel to the metal surface matches the wavevector of the surface plasmon. At these angles the excitation of surface plasmon modes reduces the reflected beam intensity creating minima in the reflectance. The measured reflectance minimum is controlled with two experimental variables, the matching fluid thickness and the angle of incidence of the light at the prism/SP structure interface. Figure 4 shows contour plots of a simulation of reflected intensity as a function of angle and oil thickness for the two structures with and without MgF$_2$. Line cuts through these simulations, identified by numbers 1-3, are shown in Fig. 5(a). Corresponding experimental reflectance curves also numbered 1-3 are shown in Fig. 5(b).

Part (a) of Figure 4 represents the reflectance contour plot from a symmetric LRSP structure (oil/Ag/Pyrex) with the oil having nearly the same index as the Pyrex. As described above, two modes, the a_b and s_b, are expected. Both are observed in the contour plot for a relatively thick oil film (300-600nm). They have been identified in the figures as the LRSP mode, corresponding to s_b, and the short range surface plasmon (SRSP) mode corresponding to a_b. In this region of thicker oil, light from the prism evanescently couples into the layered structure and the prism itself is not a large perturbation to the LRSP configuration. This changes as the oil film is compressed to below 100 nm. The structure transforms from an LRSP configuration to a (Prism/Ag/Pyrex) configuration in the limit of zero oil thickness. Now only a single SP mode is observed as a function of angle and it corresponds to a conventional surface plasmon excited at the Ag/Pyrex interface. Experimentally, it is very difficult to independently determine the oil layer thickness. However, by tuning oil thickness and looking for the transition from two SP modes to one, measurements in both regimes can be made. Curve 2 in Fig. 5(b) is a measurement at an oil thickness where both the LRSP and SRSP modes are present. The agreement with the simulation curve 2 in Fig. 5(a) in the angle at which minima occur and overall shape is very good.

Part (b) of Figure 4 shows a contour plot of calculated reflectance from an asymmetric structure (Oil/Ag/MgF$_2$/Pyrex) in which a ULRSP mode should occur. We see very similar behavior to that observed in part (a) as oil thickness is reduced with a transition from two modes (including the long range ULRSP mode of interest) to one mode occurring at around 200nm. In addition, the presence of the MgF$_2$ layer causes the ULRSP reflectance minimum to shift toward the critical angle with a narrowing of the reflectance dip (curve 2 in Fig. 5(a)). In Fig. 5(b) two measured reflectance curves are shown corresponding to a thicker oil layer (curve 2) and a thin oil layer (curve 3). Again, there is excellent agreement with the simulations in Fig. 5(a). Of particular note are the disappearance of the SRSP mode and the slight increase in angle of the ULRSP mode as oil thickness decreases and it transitions to a conventional SP mode confined primarily to the Pyrex side of the Ag. These observations confirm that the lower angle peak of curve 2 in Fig. 5(b) is the ULRSP mode predicted by the modeling work discussed above. While not shown here, we have also explored theoretically and experimentally the dependence of the ULRSP reflectance minimum on increased MgF$_2$ thickness. In agreement with simulations, the peak shifts to shorter angle and narrows as oxide thickness is increased. Cutoff occurs at approximately 300nm thickness.

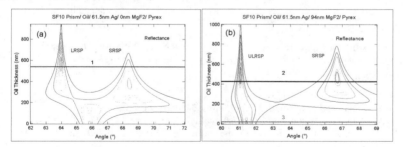

Figure 4. Reflectance contour plots as a function of oil thickness and angle of incidence for oil/Ag/MgF$_2$/Pyrex structures with 0nm MgF$_2$ (a) and 94nm MgF$_2$ (b). The horizontal lines indicate the locations of the reflectance plots in Fig. 5(a).

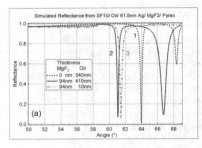

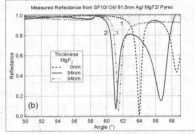

Figure 5. Simulated (a) and measured (b) reflectance for two samples with 0nm MgF$_2$ (curve 1) and 94nm MgF$_2$ (curves 2 and 3) corresponding to line cuts 1, 2 and 3 in Figure 4. Curve 1 shows the LRSP and SRSP, left and right minima, respectively. Similarly, Curve 2 shows the ULRSP and the SRSP minima. Curve 3 corresponds to a very thin oil layer. The SP mode observed here corresponds to a conventional mode propagating primarily on the MgF$_2$ side of the Ag.

CONCLUSIONS

We have explored a design that enables the implementation of low-loss, bound SP waves on metal films in MOS structures in a way that is compatible with well established silicon technology. The propagation characteristics of these modes, and their dependence on the asymmetry of the structure have been determined. In particular, we find that addition of a thin layer of low refractive index (an oxide) on one or both sides of a metal film, instead of being detrimental to SP propagation, can lead to ULRSP modes with propagation lengths beyond that of the conventional LRSP for given metal thickness. We have experimentally confirmed the existence of these ultra long range modes in a visible wavelength analog for an MOS structure with good overall agreement between modeled and measured results. Using structures that support ULRSP modes, we envision an SP modulator built around an MOS capacitor. The voltage on the capacitor would change the refractive index in a thin layer near the oxide, which would modulate the propagation of the mode. The greater propagation length of ULRSP would allow an increase in the sensitivity of the device to the small carrier-induced index changes.

ACKNOWLEDGMENTS

This work was supported by Air Force Office of Scientific Research under project FA9550-06-1-0548.

REFERENCES

1. A. Liu et al, Nature **427**, 615, (2004).
2. G. I. Stegeman and J. J. Burke, J. Appl. Phys. **54**, 4841, (1983).
3. T. Nikolagsen, K. Lesson, and S. I. Bozhevolnyi, Appl. Phys. Lett. **85**, 5833, (2004).
4. C. G. Durfee, T. E. Furtak, R. T. Collins, and R. E. Hollingsworth, Submitted to J. Appl. Phys. (2008).
5. J. Guo, and R. Adato, Opt. Exp. **14**, 12409, (2006).

Mater. Res. Soc. Symp. Proc. Vol. 1077 © 2008 Materials Research Society   1077-L07-18

Terahertz Metamaterials on Thin Silicon Nitride Membranes

Xomalin G. Peralta[1,2], C. L. Arrington[1], J. D. Williams[3], A. Strikwerda[4], R. D. Averitt[4], W. J. Padilla[5], J. F. O'Hara[6], and I. Brener[1,2]

[1]Sandia National Laboratories, P.O. Box 5800, MS1082, Albuquerque, NM, 87185
[2]CINT SNL, Albuquerque, NM, 87185
[3]University of Alabama, Huntsville, Huntsville, AL, 35899
[4]Boston University, Boston, MA, 02215
[5]Boston College, Chestnut Hill, MA, 02467
[6]MPA CINT, Los Alamos National Laboratory, Los Alamos, NM, 87545

ABSTRACT

The terahertz (THz) region of the electromagnetic spectrum holds promise for spectroscopic imaging of illicit and hazardous materials, and chemical fingerprinting using moment of inertia vibrational transitions. Passive and active devices operating at THz frequencies are currently a challenge, and a promising emerging technology for such devices is optical metamaterials. For example, a chem/bio sensing scheme based on the sensitivity of metamaterials to their dielectric environment has been proposed but may be limited due to the large concentration of electric flux in the substrate. In addition, there is an interest in fabricating 3D metamaterials, which is a challenge at these and shorter wavelengths due to fabrication constraints. In order to address both of these problems, we have developed a process to fabricate THz metamaterials on free-standing, 1 micron thick silicon nitride membranes. We will present THz transmission spectra and the corresponding simulation results for these metamaterials, comparing their performance with previously fabricated metamaterials on various thick substrates. Finally, we will present a scheme for implementing a 3D THz metamaterial based on stacking and possibly liftoff of these silicon nitride membranes.

INTRODUCTION

Metamaterials are artificial materials formed of an array of subwavelength ($\sim \lambda/10$) metallic resonators within or on a dielectric or semiconducting substrate. They exhibit electromagnetic (EM) properties not readily available in naturally occurring materials [1-3], such as negative index of refraction. In addition, their response is scalable from radio [4] to optical frequencies [5]. Therefore, they have the potential to provide a scale-invariant design paradigm to create functional materials which can enhance our ability to manipulate, control, and detect EM radiation. The recent growth in the field of metamaterials is partly due to the promise of new devices that exploit these novel EM properties in all frequency ranges, including terahertz frequencies (1 THz $\sim$ 300 µm) [6, 7]. Some of these devices require the fabrication of three-dimensional metamaterials; by stacking individual layers, by creating arbitrarily curved surfaces, or a combination of both [2].

We have developed a process to fabricate THz metamaterials on large area, free-standing thin silicon nitride (Si_3N_4) membranes. Fabricating metamaterials on thin membranes reduces any dielectric losses due to the substrate, it eliminates the asymmetry in the fringing fields and enables the implementation of various planar layering schemes. In addition, it allows us to release the membranes and wrap them over curved surfaces, therefore showing a path towards creating arbitrarily curved 3D metamaterials.

At THz frequencies, chem/bio sensing schemes have been suggested that make use of the sensitivity of metamaterials to their dielectric environment. Eliminating the substrate and the asymmetry in the fringing fields could potentially increase the cross-area of interaction with the chem/bio molecules, increasing their sensitivity.

EXPERIMENT

Fabrication

The metamaterials were fabricated on 550 μm thick 4" silicon (Si) wafers PECVD coated with 1 μm of low-stress Si_3N_4. PECVD coats all sides of the wafer, so the first processing step is to remove the Si_3N_4 from the back in four ~ (3.2 x 2.4) cm^2 areas which will be used to open windows to the metamaterials on the front side, figure 1a), left column. The windows were defined in JSR 5740 photoresist (PR) using standard photolithography before etching the Si_3N_4 using reactive ion etching (RIE) in a CF4 (40 sccm) and O_2 (10 sccm) atmosphere at 100 mTorr with a power of 100 W for 30 min. Once the Si_3N_4 is removed from the windows on the back

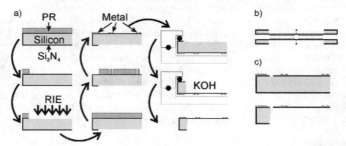

Figure 1. Schematic diagrams of a) the processing steps required to fabricate metamaterials on silicon nitride membranes, b) bilayer stacking and c) chemical liftoff process.

side, the wafer was flipped over and the metamaterials were patterned in AZ 5214 PR before evaporating 200Å of Ti followed by 500Å of Au, figure 1a), middle column. After liftoff the wafer was flipped over once more and mounted into a commercial wafer holder from AMMT GmbH that protects the front side during the next step, figure 1a), right column. A KOH bath at 30% dilution for 6 – 8 h at 80°C selectively removes the Si substrate in the Si_3N_4–free window areas defined on the back side in the first process and stops at the Si_3N_4. When removed from the wafer holder, the metamaterials are patterned onto four large-area, free-standing, thin Si_3N_4 windows, see figure 2. Each window has four different metamaterial designs, each covering a (1.3 X 1) cm^2 area. One section in one window is intentionally left blank to use as a reference.

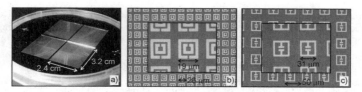

Figure 2. a) Fully processed wafer with the four Si_3N_4 windows evident, b) double Split-Ring Resonator (dSRR) array and b) Electric resonator (E2) array.

Characterization of planar metamaterials on Si_3N_4 membranes

In order to assess the effect of having a thin (1 μm) Si_3N_4 membrane instead of a thick substrate (hundreds of microns) as substrate for THz metamaterials, we characterized the electromagnetic response of all 15 metamaterials with a THz time-domain spectroscopy (THz-TDS) system. The THz-TDS system is based on photolithographically defined photoconductive antennas for both the source and the detector and has been detailed elsewhere [8]. The experiments were performed at room temperature in a dry (<1% humidity) air atmosphere. The THz beam diameter is ~ 3 mm and is easily contained within the sections covered by a particular metamaterial design. The time-varying electric field of the THz transmitted light through the blank Si_3N_4 membrane and through the metamaterials were recorded. After a numerical Fourier transformation, the THz transmission spectra and the phase change relative to the reference were obtained. The THz radiation was polarized perpendicular to the gaps and transmitted normally through the plane of the metamaterials, see figure 3.

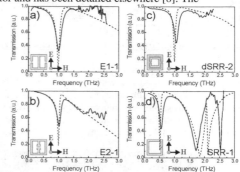

Figure 3. Transmission spectra (solid) and simulation results (dash) for a) E1, b) E2, c) dSRR and d) SRR on planar membranes. Labels refer to designs in Table I.

Electromagnetic modeling was also performed using finite-element software [9]. A constant dielectric permittivity of 7 was used for Si_3N_4 in the simulations. Figure 3 shows the transmission spectra and simulation results for 4 representative metamaterial designs. In most cases we were able to reproduce qualitatively the features observed in the experimental transmission spectra and their relative positions.

Table I lists geometrical parameters and quality factors (Q) of some of the resonators fabricated on thin Si_3N_4 membranes (in bold) as well as physically similar ones found in the literature fabricated on thick substrates. We calculated $Q = \omega_0 / \Delta\omega$, where ω_0 is the resonant frequency and $\Delta\omega$ is the full width at half maximum, for all low-frequency resonances. The first 7 samples where chosen because the geometrical parameters are very similar and the last 4 have very similar low-frequency resonance positions (0.9 THz dSRR in [12] and 0.5 THz SRR in [7]). The geometrical parameters are g – gap, w – metal linewidth, s – separation between rings (dSRR) or capacitor plate width (E2), l – outer dimension, p – lattice constant.

Table I. Geometrical parameters and quality factors. All lengths are in μm.

Design	Ref.	g	w	s	l	p	Substrate	Thickness	Metal	Thickness	Q
E1	10	2	4		36	50	SI GaAs	670 μm	Ti/Au	210 nm	6.0
E1-1		2	4		**36**	**50**	Si$_3$N$_4$	1 μm	Ti/Au	70 nm	4.5
E2	10	2	4		36	50	SI GaAs	670 μm	Ti/Au	210 nm	7.0
E2-1		2	4	18	**38**	**50**	Si$_3$N$_4$	1 μm	Ti/Au	70 nm	**6.4**
dSRR	11	2	6	3	36	50	Silicon	640 μm	Al	200 nm	7.6
	11	2	6	3	36	50	Quartz	1.03 mm	Al	200 nm	8.0
dSRR-1		2	6	3	**38**	**52**	Si$_3$N$_4$	1 μm	Ti/Au	70 nm	**7.3**
	12	4	4	3	30	44	Quartz	0.8 mm	Cu	320 nm	6.3
dSRR-2		2	4	4	**29**	**46**	Si$_3$N$_4$	1 μm	Ti/Au	70 nm	**6.3**
SRR	7	2	6		36	50	HR GaAs	670 μm	Cu	3 μm	6.2
SRR-1		2	4		**55**	**75**	Si$_3$N$_4$	1 μm	Ti/Au	70 nm	**4.4**

Three-dimensional implementations and characterization

Our first approach to implement a 3D metamaterial using thin Si$_3$N$_4$ membranes is to fabricate two wafers with the same metamaterial patterns and stacking them with the metamaterials facing each other, figure 1b). This opens up the possibility of studying the metamaterial properties as a function of separation between the layers, registry between the units and relative orientation of the units [13, 14]. To avoid artifacts in the measurements, the wafers must be kept parallel by introducing a spacer of a known thickness around the edges.

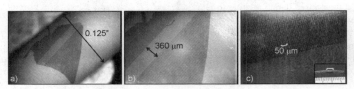

Figure 4. Microscope images of a piece of metamaterial covered, Si$_3$N$_4$ membrane, wrapped around a) and b) a Teflon tube and c) a 1.5 mm diameter tubing. Scale of inset mm.

Another approach is to remove the patterned membrane from the Si wafer after fabrication using a sharp razor blade. The membrane can then be placed onto a host material which can be planar, such as a piece of mylar of a known thickness for the stacking scheme, or a 3D object such as a Teflon tube (mylar and Teflon® PTFE are transparent at THz frequencies). Figure 4 shows two different instances of a curved metamaterial fabricated in this manner.

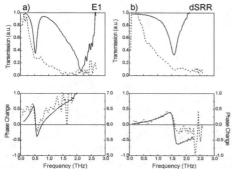

Figure 5. Transmission spectra (top) and phase change (bottom) for a) E1 and b) dSRR on planar (solid) and curved (dash) membranes.

Figures 5 illustrates the frequency-dependent amplitude transmission and the phase change of two different metamaterial designs wrapped around a Teflon tube, fully covering the side exposed to the THz beam. The reference was a bare piece of Teflon tube. For comparison, they are plotted with the corresponding planar metamaterial covered membrane's response. In both curved metamaterials we observe a small decrease in transmission at the resonance of the planar metamaterials on a uniformly decreasing background. The phase change data confirms this observation.

DISCUSSION

We successfully fabricated and characterized THz metamaterials on 1 μm thick Si_3N_4 membranes. The difference between transmission measurements and simulations may be related to having used a constant dielectric permittivity ε in the modeling instead of $\varepsilon(\omega)$ (unavailable in the THz region). The largest difference occurred for the high frequency resonances in SRR-1 (figure 3d). This might be due to the effect of higher-order modes [15] but it is still under investigation.

All of the structures fabricated on Si_3N_4 membranes have comparable, although slightly smaller, Q's to similar structures that appear in the literature. To understand this result we need to consider the role of the substrate, or lack thereof, the effect of coupling between resonators due to a modified fringing fields as well as the metallization thickness (20 nm Ti/ 50 nm Au) which is less than the skin depth (δ) of the THz radiation, at 1 THz, $\delta_{Au} \sim 75$ nm, $\delta_{Ti} \sim 325$ nm. It is interesting to note that the resonances are better defined than the only reference in the literature to THz metamaterials on thin silicon nitride membranes [16].

The approach described for layering is currently limited to two layers due to the thickness of the Si wafers. To extend this stacking scheme to more layers, we can reduce this thickness, use smaller pieces which can be nested into each other, use dielectric spacers or remove the membranes from the substrate. Currently the liftoff procedure for the Si_3N_4 membrane is done manually, therefore it is not well controlled and tearing occurs. To overcome this problem we can define a Si_3N_4–free trench at the edge of the window area covered by the metamaterials on the front side **before** actually patterning the metamaterials. When the wafer is placed in KOH, the membrane will be released, figure 1c).

To our knowledge, this is the first implementation and measurement of a curved metamaterial at THz frequencies. The resonances in transmission can be resolved although riding on top of a currently unexplained, uniformly decreasing background, but are clearly defined in the phase change data. Two future experiments on planar metamaterials which can help understand this response are: 1) Study the transmission as a function of angle between the incident field and the plane of the metamaterials as having a curved surface is equivalent to multiple angles simultaneously. 2) Look at the effect of having a Teflon (dielectric) backing. Clearly, further studies need to be done to fully understand the response.

CONCLUSIONS

We have successfully fabricated THz metamaterials on large-area, free-standing, thin Si_3N_4 membranes and have demonstrated one scheme for fabricating layered and curved metamaterials. The performance of the planar metamaterials on membranes, as judged by the Q of the low-frequency resonances, is comparable to those fabricated on thick substrates. A better understanding of the role of the effective permittivity of the supporting media and metallization thickness might help improve their response. We successfully characterized the first implementations of curved THz metamaterials and corroborated the location of their resonant response, but further studies are needed to fully understand the response. Metamaterials on thin membranes may prove practical for improving the sensitivity of metamaterial based chem/bio sensors and implementing truly 3D metamaterial structures across the electromagnetic spectrum.

We would like to thank A. K. Azad for help with the measurements on the 3D structures. We acknowledge support from the CINT and the IC Postdoctoral Research Fellowship Program (XGP). Sandia is a multiprogram laboratory operated by Sandia Corporation, a Lockheed Martin Company, for the US DOE's NNSA under Contract DE-AC04-94AL85000.

REFERENCES

1. D. R. Smith, W. J. Padilla, D. C. Vier, S. C. Nemat-Nasser and S. Schultz, *PRL* **84**, 4184 (2000).
2. R. A. Shelby, D. R. Smith and S. Schultz, *Science* **292**, 77 (2001).
3. J. B. Pendry, A. J. Holden, D. J. Robbins and W. J. Stewart, *IEEE Trans. Microwave Theory Tech.* **47**, 2075 (1999).
4. M. C. K. Witshire, J. B. Pendry, I. R. Young, D. J. Larkman, D. J. Gilderdale and J. V. Hajnal, *Science* **291**, 849 (2001).
5. C. Enrich, M. Wegener, S. Linden, S. Burger, L. Zschiedrich, F.Schmidt, J. F. Zhou, Th. Koschny and C. M. Soukoulis, *PRL* **95**, 203901 (2005).
6. B. Ferguson and X-C Zhang, *Nature Materials* **1**, 26 (2002).
7. W. J. Padilla, A. J. Taylor, C. Highstrete, M. Lee and R. D. Averitt, *PRL* **96**, 107401 (2006).
8. J. F. O'Hara, J. M. O. Zide, A. C. Gossard, A. J. Taylor and R. D. Averitt, *APL* **88**, 25, 251119 (2006).
9. CST Microwave Studio ®, © 2005 CST – Computer Simulation Technology, Wellesley Hills, MA, USA. www.cst.com
10. W. J. Padilla, M. T. Aronsson, C. Highstrete, M. Lee, A. J. Taylor and R. D. Averitt, *PRB* **75**, 041102R (2007).
11. A. K. Azad, J. Dai and W. Zhang, *Opt. Lett.* **31**, 5, 634 (2006).
12. X.-L. Xu, B.-G. Quan, C.-Z. Gu and L. Wang, *J. Opt. Soc. Am. B* **23**, 6, 1174 (2006).
13. P. Gay-Balmaz and O. J. F. Martin, *J. Appl. Phys.* **92**, 2929 (2002).
14. N. Katsarakis, G. Konstantinidis, A. Kostopoulos, R. S. Penciu, T. F. Gundogdu, M. Kafesaki, E. N. Economou, Th. Koschny and C. M. Soukoulis, *Opt. Lett.* **30**, 1348 (2005).

15. J. F. O'Hara, E. Smirnova, H.-T. Chen, A. J. Taylor, R. D. Averitt, C. Highstrete, M. Lee and W. J. Padilla, *J. Nanoeleco. Optoelectron.* **2**, 90 (2007).
16. M. C. Martin, Z. Hao, A. Liddle, E. H. Anderson, W. J. Padilla, D. Schurig and D. R. Smith, *Conference Proceedings of IEEE IRMMW-THz 2005*, vol. **1**, 34 – 35 (2005).

Mater. Res. Soc. Symp. Proc. Vol. 1077 © 2008 Materials Research Society   1077-L07-20

Effect of Ag and Au Doping on the Photocatalytic Activity of TiO2 Supported on Textile Fibres

Mohammed Jasim Uddin, Federico Cesano, Domenica Scarano, Silvia Bordiga, and Adriano Zecchina
Department of Chemistry IFM and Nanostructured Interfaces and Surfaces (NIS), Centre of Excellence, University of Turin, Via P. Giuria 7, Turin, 10125, Italy

ABSTRACT

A simple method to develop TiO_2, Ag or Au-doped TiO_2 thin films on cotton textiles for advanced applications, is reported. The homogeneous TiO_2 thin films have been deposited on cotton textiles by using sol-gel method at low temperature (100° C), whereas Ag and Au nanoparticles were then deposited on the pre-existent TiO_2 films by photoreduction. The Ag/TiO2 covered cotton fibres show multichromic behaviour (grey colour under visible light and brown colour upon ultraviolet light exposure) as well as photoactivity. The Au/TiO2 film coated the cotton textile produces a purple colour with excellent self cleaning properties. The original and treated fibres have been characterized by several techniques (SEM, HRTEM, FTIR, Raman, UV–vis spectroscopy and XRD).

INTRODUCTION

Nanosized anatase TiO_2 is known to have a wide range of applications in photocatalysis[1]. The optical properties and the photocatalytic activity of TiO_2 coatings do not depend only on the phase, but also on the crystallite size and porosity. The only drawback of TiO_2 is that its band gap lies in the near-UV of the electromagnetic spectrum: 3.2 eV (285 nm) and 3.0 eV (410 nm) for anatase and rutile, respectively. As a consequence, only UV light is able to create electron-hole pairs and to initiate the photocatalytic processes. It is therefore evident that any modification of the TiO_2-based photocatalysts, resulting in a lowering of its band gap or in the introduction of stable optical sensitizers, is representing a breakthrough in the field[2]. An exhaustive analysis of the different approaches used to dope TiO_2 is beyond of scope of this contribution. One case is doping TiO_2 with various transitions metals such as Au, Ag, Pt, Cr, Nb, V, Mn, and Fe[3-8]. Brook et al.[9] have described that Ag/TiO2 films are not only active as disinfectants but also exhibit "self-regeneration" capability, because they both kill bacteria present on the film surface and photo-degrade the residues. Recently, it has been shown that Ag/TiO2 films on pyrex glass show also multicolor photochromism and photoinduced conversion of Ag nanoparticles due to surface plasmon resonance effects[10]. A maximum in the photocatalytic activity with the Au-modified samples has been registered in case of 0.16 wt.% Au/TiO2. At this content the activity of the Au-modified TiO_2 is approximately double that of the semiconducting support[11, 12]. For the reasons briefly mentioned above, we became interested in modifying the light absorbing property and antibacterial activity of Au/TiO2 and Ag/TiO2 films covering the cotton fabrics. To check the reuse of the Ag/TiO2 and Au/TiO2 covered fibres, multiple adsorption–photodegradation cycles have been performed.

EXPERIMENTAL DETAILS

Pure cotton fibre, 10-15 μm in diameter, from PVS srl (Milano, Italy) was used for the entire process. All chemicals used in this work were procured from Aldrich, Germany, and have been used as received. The procedure for the formation of Ag/TiO2 and Au/TiO2 coated cotton fibres consists of several stages which are described in references[13-15]. The resulting sol was transparent and quite stable and can be used to impregnate cotton fibre. In addition, the resultant samples were dried in a preheated oven at 50° C and were separately soaked in 0.001M AgNO$_3$ and 0.001M HAuCl$_4$ aqueous solution for 1 minute (two different samples). The samples were then dried at room temperature. As prepared samples were irradiated at 308 K (50 mW/cm^2, approx 295 nm-3000 nm, Polymer GN 400 ZS, Helios Italquartz, Italy) for 15 minutes (AgNO$_3$ soaked sample) and for 30 minutes (HAuCl$_4$ soaked sample) in air at atmospheric humidity.

The morphology of pure cotton fibre and cotton fibre supported Ag/TiO2 film were studied by Scanning Electron Microscopy (Leica, Stereoscan 420) equipped with EDX microanalysis system (Oxford). The UV-vis reflectance spectra were obtained at room temperature on a Perkin-Elmer UV-Vis-NIR, to investigate the UV absorption properties, the location of the absorption edge and the quantum size effects (if any) of the synthesized film. XRD patterns have been collected by means of Philips PW1830 x-ray diffractometer in a Bragg Brentano configuration to identify the crystal phase and the structure.

RESULTS AND DISCUSSION

Thin film morphologies and XRD analyses

From SEM images, shown in Fig. 1, the surface morphology of the pure and sol-gel treated cotton fibres is compared. The enlarged view of Ag/TiO2 and Au/TiO2 covered fibres are shown in Fig 1(c and d). As the presence of Ag/TiO2 or Au/TiO2 coating clearly leads to filling of the folds characteristic of the virgin fibre, the formed coating appears homogeneously deposited on the fibre surface. The formation of firmly grafted films is likely associated with the high number of hydrophilic groups present on the cotton fibre, which are preferential anchoring sites for the deposition of TiO$_2$. To quantify the amount of TiO$_2$ present on the fibre, one of the simplest method adopted was to burn the organic support and to weight the final inorganic residue. The resulting figure was in the 6-7 wt % range. However, the most interesting result, coming from this procedure, became evident from SEM analysis of the morphology of the hollow shape inorganic residue. In Fig. 1 (e and f) the morphology of Ag/TiO$_2$ films after burning at 500°C for 5h is reported. From this figure it is evident that the coating preserves the morphology of the original film. Due to the absence of the material inside, after burning, the film structure is partially collapsed and damaged during the manipulation. From the film edges, the thickness of the layer can be inferred: the arrows indicate a thickness of ~ 95-100 nm (inset of Fig 1e). The film thickness, however, is sufficient to cover the natural folds of the fibres present on the surface. The silver nanoparticles (NPs) dispersed on the TiO2 film are shown in Fig. 1c (indicated by white arrow) and also confirmed by AFM analyses (figure not shown for the sake of brevity). From the secondary backscattered electron image (figure not shown for the sake of brevity) and the EDX analysis, it is confirmed that the white spots shown in the SEM images (Fig. 1d) are indeed gold particles. The EDX analysis of Ag/TiO2 and Au/TiO$_2$ covered cotton fibres after burning at 500 °C, are compared in Fig. 2a where the presence of Ag in Ag/TiO$_2$ and Au in Au/TiO$_2$ is evidenced well. The quantitative analyses of the photoactive thin film indicate that the elemental composition of silver in Ag/TiO$_2$ and gold in Au/TiO$_2$ are 0.3 and 0.1 %

(w/w) respectively (Fig. 2a). The BET results of the residual samples indicate that the film consists of porous materials of about 15 m²/g surface area.

The XRD patterns of the pure, TiO₂, Ag/TiO2 and of Au/TiO2 coated cotton fibre are illustrated in Fig. 2b. In the inset, an exploded view of the patterns of the Ag/TiO2 and Au/TiO2 impregnated sample are shown together with the position of the peaks of anatase taken from the reference. The major peak at 26.4° is due to the crystalline phase of cotton fibres[16]. The three broader peaks at 29.44°, 44.192° and 56.40° are indicative of anatase phase. The remarkable width of these peaks suggests that the particles sizes of the deposited photocatalyst are quite small. From the full width at half maximum (FWHM) of the peaks at 29.44⁰ and 56.40⁰ and by using Scherrer's equation, $L_c = K\lambda/(\beta\cos\theta)$[17] (where λ is the X-ray wavelength, β is the FWHM

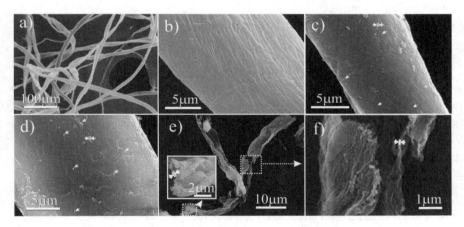

Figure 1: SEM images of: a) a tangle of pure cotton fibres; b) an enlarged view of a pure fibre shown in a), where folds quite parallel to the fibre-axis are illustrated; c) and d) cotton fibres coated with films of Ag/TiO₂ and Au/TiO₂, respectively. The particles, evidenced by arrows in these two figures, correspond to Ag or Au nanoparticles with size larger than 50 nm. The majority of the small metal particles cannot be evidenced by SEM. e) a Ag/TiO2 film cotton fibre sample burned at 500°C; f) an enlarged view of a portion shown in e).

of the diffraction line, θ is the diffraction angle, and K is a constant, which has been assumed to be 0.9), an average particles diameter of silver and gold doped TiO₂ is estimated to be about 3.5 nm and 3.0 nm respectively, while that of the TiO₂ particles on the bare TiO₂ film is about 5.0 nm. An additional peak at $2\theta = 44.5°$, obtained on the pattern of the Au/TiO2 covered cotton fibre, is corresponding to Au° particles on the TiO₂ film. From the width of the $2\theta = 44.5°$ peak, the size of the Au particles is estimated to be ~40 nm. This figure is different from the average particle size estimated in TEM analysis. We believe that this signal is due to the few large Au particles imaged with TEM *(figure not shown for the sake of brevity)* and SEM analyses. In fact due to their large size, these particles are expected to give a substantial contribution to the XRD pattern. On the contrary, the silver nanoparticles dispersed on the TiO2 phase are not detectable at this composition. The photochromic contribution is only due to a little quantity of silver nanoparticles homogeneously incorporated in TiO2 film without forming any large particles that may contribute significantly to XRD pattern. We have observed that the XRD results are in strong agreement with the UV-vis and Raman results *(figure not shown for the sake of brevity)*.

UV-vis reflectance spectra, photochromic effect and photocatalytic self cleaning

The absorption edge at ~ 29000 cm^{-1}, observed for TiO$_2$ and Ag/TiO2 coated cotton fibre (Fig. 3 curves 3 and 4), is very similar to that of pure anatase TiO$_2$ (curve 2). The slightly upward shifted absorption edge suggests that the particles sizes of the photocatalyst are smaller than those of anatase (quantum size effect). Coming now to curve 4, the presence of a complex absorption centered at ~ 22000 cm^{-1} is strongly indicating the presence of Ag particles on the TiO$_2$ film (since Ag$^+$ does not have absorption in this range).

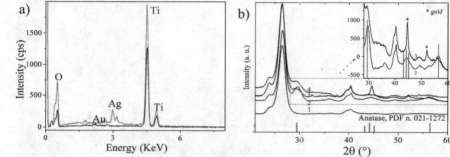

Figure 2: a) EDX spectra of the Ag/TiO2 (gray line) and of Au/TiO2 coated cotton fibres (black line); b) XRD pattern of: pure cotton fibres (curve 1), TiO$_2$-coated cotton fibres (curve 2), Ag- and Au/TiO2 coated cotton fibres (curve 3 and 4, respectively). The vertical lines, representing the peak positions of the standard anatase, are reported below the XRD pattern. The most intense XRD diffraction peaks are related to cellulose. In the inset, the most representative portion of the pattern, showing the TiO$_2$ anatase phases, is reported.

Upon exposure to UV light, the intensity of the ~22000 cm^{-1} absorption (showing a tail extending in the NIR) is strongly enhanced. This is due to the reduction of Ag$^+$ to Ag° and formation of (Ag°)$_n$ clusters. The color of the sample changes from light grey to brown. Furthermore, it is known that Ag particles deposited on TiO$_2$ are responsible for multi colour photochromic behaviour. This is explained with the anisotropic character of Ag nanoparticles, which have two different resonance wavelengths in the porous TiO$_2$ film[10]. It is known that, after photon absorption, the presence of silver nanoparticles promotes the charge separation of the electron–hole pairs from TiO$_2$, by acting as an electron sink[9]. Kato et al.[18] explained that photodeposition of Ag on a TiO$_2$ film enhanced photocatalytic degradation of gaseous sulphur compounds and suggested that Ag acted as a co-catalyst. In Fig. 3(b), a sequence of light exposure cycles to visible and ultraviolet irradiation is shown. This sequence causes the fibres to become alternatively brown under ultraviolet light and white grey under visible light and this behaviour can be repeated several times without substantial modification. Matsubara et al.[19] described the tendency of the Ag NPs to become flattened with increasing irradiation time and the lateral diameter to increase from 15 to 35 nm. This reason plausibly can be responsible for the slight overall

156

decrease of the reversible photochromism. On the contrary, the Au/TiO2 films do not display any photochromic behaviour because of the excellent nobility of gold. The photoactivity of the gold-titanium dioxide coated cotton textiles has been investigated by exposing the samples containing adsorbed methylene blue (MB), an organic stains used to make dirty the cotton textiles, to solar-like light. The UV–vis reflectance spectra before (curve a) and after illumination with increasing time obtained on Au/TiO$_2$-cotton/MB systems are compared in Fig. 3c (inset). From this comparison it is inferred that the complex absorption bands in the 20000–12000 cm^{-1} interval, due to absorption of MB, change rapidly because Au/TiO2 promotes the catalytic photodegradation. Of course the disappearance rate of the band due to adsorbed MB on the TiO$_2$-covered fibres is much higher than that observed in case of untreated fibres. An important result of this investigation is the observation that the TiO$_2$ covered cotton fibres containing silver and gold particles are more active in MB degradation than those not containing gold (Fig 3d). In particular, the dispersion of gold nanoparticles on the TiO$_2$ matrix has a beneficial influence on the organic stains photodegradation and catalysis[12, 20].

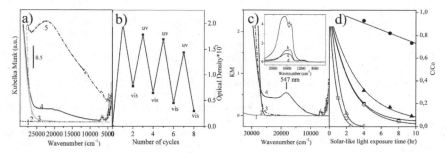

Figure 3: (a) UV-vis reflectance spectra of: fibres treated with blank sol (1); anatase (2); TiO$_2$-coated cotton fibres (3); Ag/TiO2 coated cotton fibres under visible light (4) and UV light exposure (5), respectively. This absorption band is related to the different optical properties of the Ag/TiO2 coating under alternative UV/Vis light exposures. (b) Optical density of the absorption band between 26750 and 7000 cm^{-1}. The sequence of the light exposure cycles represents the alternative and reversible optical behaviour after ultraviolet (maxima) and visible (minima) light irradiations. (c) UV-vis reflectance spectra of: pure cotton fibre (1), anatase (2), TiO$_2$ (3) and Au/TiO2 (4) covered cotton fibre, respectively. The inset represent the MB degradation on Au/TiO2-cotton upon solar light exposure with time (curve b-d) (d) MB Photodegradation upon solar like light illumination of: pristine fibres (●), TiO$_2$-coated fibres (►), Au/TiO2-coated fibres (■), Ag/TiO2-coated fibres (○), P25 (△) and Au-P25 (□).

Figure 4 shows that the photocatalytic efficiency of the Au/TiO2 coated fibre on degrading MB is unchanged upon repeating cycles. These results suggest that the activity of Au/TiO2 film is not influenced by successive impregnation steps and this is in agreement with the results shown in our previous washing cycles study[13, 14].

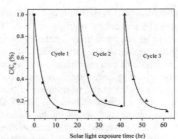

Figure 4: Photocatalytic activity of Au/TiO2 covered cotton fibre upon repeated MB adsorption–illumination cycles.

CONCLUSION

We have developed a new and simple process to deposit silver and gold doped TiO_2 for efficient self cleaning, photochromic and antibacterial application. The morphological studies and the characterization results indicate that the deposition of Ag/TiO2 and Au/TiO2 coating on the cotton fibres and the curing at low temperature, have not changed the basic chemical and physical properties of the cotton fibres. In addition, we can conclude that the Ag–TiO_2 thin film will act as a photoactive materials of multi functionality. Furthermore, we expect that the film could display biocidal functionality to bacteria, phtochtomism (Ag particles) and photoactivity for self cleaning (TiO_2 phase). Cotton fibres, covered by a thin Au/TiO2 film constituted by anatase nano crystallites strongly adhering to the support, display a purple color and are photocatalytically active under solar light. The Au/TiO2 covered cotton fibres show high efficiency in MB photodegradation, so suggesting high photocatalytic self cleaning properties under solar like light.

REFERENCES

1   A. Fujishima and K. Honda, *Nature*, 1972, **238**, 37 - 38.
2   S. Usseglio, A. Damin, D. Scarano, S. Bordiga, A. Zecchina and C. Lamberti, *J. Am. Chem. Soc.*, 2007, **129**, 2822 - 2828.
3   M. Anpo and M. Takeuchi, *J. Catal.*, 2003, **216**, 505-516.
4   O. Carp, C. L. Huisman and A. Reller, *Prog. Solid State Chem.*, 2004, **32**, 33-177.
5   W. Y. Choi, A. Termin and M. R. Hoffmann, *J. Phys. Chem.*, 1994, **98**, 13669-13679.
6   C.-G. Wu, C.-C. Chao and F.-T. Kuo, *Catal. Today*, 2004, **97**, 103-112.
7   S. Rodrigues, K. T. Ranjit, S. Uma, I. N. Martyanov and K. J. Klabunde, *J. Adv. Mater.*, 2005, **17**, 2467-2471.
8   P. N. Kapoor, S. Uma, S. Rodriguez and K. J. Klabunde, *J. Mol. Catal.*, A, 2005, **229**, 145-150.
9   L. A. Brook, P. Evans, H. A. Foster, M. E. Pemble, A. Steele, D. W. Sheel and H. M. Yates, *J. Photochem. Photobiol. A-Chem.*, 2007, **187**, 53-63.
10   Y. Ohko, T. Tatsuma, T. Fujii, K. Naoi, C. Niwa, Y. Kubota and A. Fujishima, *Nat. Mater.*, 2003, **2**, 29-31.

11   G. Mele, G. Ciccarella, G. Vasapollo, E. Garcia-Lopez, L. Palmisano and M. Schiavello, *Appl. Catal. B-Environ.*, 2002, **38**, 309-319.
12   A. Corma, P. Serna and H. Garcia, *J. Am. Chem. Soc.*, 2007, **129**, 6358 -6359.
13   M. J. Uddin, F. Cesano, F. Bonino, S. Bordiga, G. Spoto, D. Scarano and A. Zecchina, *J. Photochem. Photobiol. A-Chem.*, 2007, **189**, 286–294.
14   M. J. Uddin, F. Cesano, D. Scarano, F. Bonino, G. Agostini, G. Spoto, S. Bordiga and A. Zecchina, *J. Photochem. Photobiol. A-Chem.*, 2008 (in press).
15   M. J. Uddin, F. Cesano, S. Bertarione, F. Bonino, S. Bordiga, D. Scarano and A. Zecchina, *J. Photochem. Photobiol. A-Chem.*, 2008, **196**, 165-173.
16   M. A. Moharram, T. Z. A. E. Nasr and N. A. Hakeem, *J of Pol Sc.: Pol Lett Ed*, 1981, **19**, 183-187.
17   L. S. Birks and H. Friedman, *J. App. Phy*, 1946, **17**, 687-692.
18   S. Kato, Y. Hirano, M. Iwata, T. Sano, K. Takeuchi and S. Matsuzawa, *Appl. Catal. B-Environ.*, 2005, **57**, 109-115.
19   K. Matsubara and T. Tatsuma, *Adv. Mater.*, 2007, **19**, 2802-+.
20   A. Corma and P. Serna, *Science*, 2006, **313**, 332-334.

Mater. Res. Soc. Symp. Proc. Vol. 1077 © 2008 Materials Research Society

Structural Dynamics of a Single Photoreceptor Protein Molecule Monitored With Surface-Enhanced Raman Scattering Substrates

Kushagra Singhal[1], Karthik Bhatt[1], Zhouyang Kang[2], Wouter Hoff[3], Aihua Xie[2], and A. Kaan Kalkan[1]

[1]Functional Nanomaterials Lab, Department of Mechanical and Aerospace Engineering, Oklahoma State University, 218 Engineering North, Stillwater, OK, 74078

[2]Department of Physics, Oklahoma State University, 145 Physical Sciences, Stillwater, OK, 74078

[3]Department of Microbiology and Molecular Genetics, Oklahoma State University, 307 Life Sciences East, Stillwater, OK, 74078

ABSTRACT

Photoactive yellow protein (PYP) is a small cytosolic photoreceptor that actuates the negative phototactic response in its host organism *Halorhodospira halophila*. It has an optical absorption maximum at 446 nm (blue light). We report an initial study of the photocycle of PYP at the single molecule level using "high enhancement factor" surface-enhanced Raman scattering (SERS)-active nanostructures with 514 nm laser excitation. The SERS-active "nanometal-on-semiconductor" structures are prepared employing a redox technique on thin germanium films, coated on glass slides. Single molecule spectra are observed in terms of sudden appearance of discernable Raman peaks with spectral fluctuations. The single molecule spectra capture protonation, photo-isomerization, and H-bond breaking - the steps that are instrumental in the photocycle of PYP. This is indicative of single PYP molecules diffusing to high-enhancement-factor SERS sites, and undergoing photo-cycle under 514 nm excitation.

INTRODUCTION

Photoactive yellow protein (PYP) is a small (14 kDa) cytosolic photoreceptor protein with 125 amino acid residues [1]. It is responsible for the negative phototactic response of its host organism *Halorhodospira halophila* (thus the wild type PYP is called Hal-PYP) [2]. It has a *para*-coumaric acid (*p*CA) chromophore [3, 4] (Figure 1) covalently bound to side chain of Cys69 through a thiolester linkage. The photocycle of PYP is initiated by the absorption of a blue photon (absorption peak at 446 nm), whose energy thereafter thermalizes through a chain of conformational states for both the chromophore and the protein, as seen in Figure 2 [5, 6].

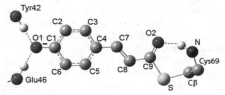

Figure 1. Structure of the pCA chromophore of PYP (pG state): 3 key active site H-bonds indicated as dotted lines.

In the initial dark (receiver or **pG**) state, the pCA is in trans configuration about the vinyl C7=C8 bond with a deprotonated phenolic oxygen [4, 7]. The carbonyl oxygen is H-bonded to the amide group of Cys69 residue, and phenolic O$^-$ is stabilized by H-bonding network involving Tyr42 and protonated [8] side chain of Glu46 [9]. Upon excitation, the chromophore photo-isomerizes to cis configuration about vinyl C7=C8 bond. Concurrently, the H-bond with the amide group of Cys69 is broken resulting in [8, 10, 11] the red-shifted intermediate (**pR**) state, which has an absorption peak at 465 nm. Proton transfer from Glu46 to phenolic O$^-$ [8, 11] leads to blue shifted **pB'** state. Subsequent re-establishment of H-bond between pCA carbonyl O and amide group of Cys69 [10] and unfolding [12] of protein molecule results in the signaling (**pB**) state, which has an absorbance maximum near 355 nm. This is the most stable state in the photocycle, and it takes 350 ms to thermally decay to the initial pG state [1, 5, 6].

While Raman spectroscopy, in itself, is a valuable technique to study the conformational dynamics of molecules through vibrational modes, the key advantage of SERS is the dramatic gains in the Raman signal intensity as high as 10^{14}-10^{15}, allowing the detection of Raman spectra from single molecules using data collection times of less than 1 s [13]. Compared to other single molecule techniques, Raman spectra contain a high degree of structural information on the pCA. Thus, SERS is capable of revealing specific structural changes at the active site of PYP at the single molecule level.

The literature reports various analytical studies on PYP using NMR, X-ray crystallography, infrared spectroscopy, Raman scattering, fluorescence spectroscopy, and time resolved Raman scattering [14, 15]. However, there are no reports of SERS studies on PYP. We have used SERS to conduct single molecule studies at low (~10^{-9} M) concentrations of PYP, using indigenously prepared metal-nanoparticle-on-semiconductor-thin-film substrates. The temporal appearance of SERS peaks with significant peak-narrowing and spectral shifts in consecutive scans (within 1 s of each other) suggest single PYP molecules finding high enhancement sites ("hot spots") on our SERS substrates excited by 514 nm-laser. In the SERS literature, this enhancement is mainly attributed to localization of electromagnetic fields in the vicinity of nanoparticles as a result of resonant coupling between coherent electron oscillations and the photons (i.e., localized surface plasmon polaritons) [13, 16-18]. This localization can be further enhanced by plasmon hybridization between adjacent nanoparticles, such that Raman scattering is detectable from single molecules [16-18].

EXPERIMENTAL DETAILS

The SERS substrates were prepared by depositing ultra-thin Ge films (~4.5 nm thick) on Corning 1737 code glass slides using a Cressington 208 High Vacuum Turbo Carbon Coater. The films were then immersed in 0.002 M AgNO$_3$ for 30 s. Ge reduces Ag$^+$ to Ag nanoparticles on the surface, thus producing the SERS-active substrates. The salient features of this technique are: no surfactants used; no capping agent used; and, size control [19]. Figure 3 shows a substrate thus prepared with an average nanoparticle size of 30 nm.

Histidine-tagged wild type PYP was prepared at a concentration of ~10 µM [8]. SERS was conducted by spotting 1 µL drops of diluted PYP (10^{-9} M) on the substrates and subsequently collecting the back-scattered radiation from the solution/nanoparticle interface with a 20× objective lens. For spectral acquisition, a Renishaw RM 1000 system was employed. SERS was excited with a Spectra-Physics 160-series 514 nm Ar$^+$ ion laser. This yields pre-resonance Raman spectra of the pCA chromophore in PYP. An integration time of 3 s, 8.3 mW incident laser power, and ~20 µm laser spot size were used to scan for ensemble-averaged SERS.

An integration time of 0.25 s, 0.2-0.4 mW laser power, and laser spot size of ~5 µm were employed for single-molecule scans.

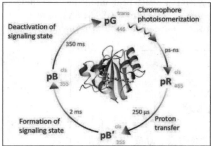

Figure 2. Photocycle model of PYP.

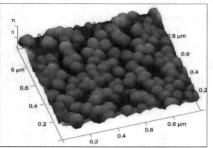

Figure 3. Representative AFM image of Ag nanoparticles reduced on Ge film.

RESULTS AND DISCUSSION

Figure 4 shows the ensemble-averaged SERS spectrum of PYP. The positions of the peaks were compared to those published for PYP. The ensemble average SERS spectrum of PYP is not identical to that published for the resonance Raman spectrum of the pG state. However, the peaks at 1162, 1281, 1480, 1550 cm^{-1} appear to correspond to peaks observed in the Raman spectrum of the pG state at 1163, 1288, 1495, and 1555 cm^{-1} [20]. Similarly, the peak at 1582 cm^{-1} seems to correspond to the reported signaling state Raman peak at 1576 cm^{-1}. Strong peaks in the PYP SERS spectrum at 1234, 1345, 1370, 1418, and 1623 cm^{-1} were not observed in the resonance Raman spectrum. This suggests that polarization effects or altered selection rules play a significant role in determining the SERS spectra of PYP.

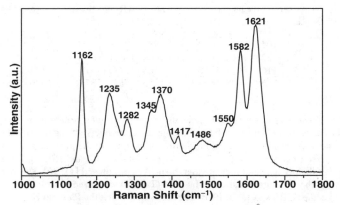

Figure 4. Ensemble-averaged SERS spectrum of 10^{-9} M PYP.

The structural dynamics of PYP was explored by capturing sudden appearance of discernable Raman peaks, as seen in Figure 5, which is indicative of a single molecule diffusing to a high-enhancement factor SERS site [18]. Comparison of SERS peak intensities for single PYP molecules with those for regular Raman scattering (obtained from 10^{-4} M PYP in the probed volume under 20× objective lens) reveals SERS enhancement factors as high as 3×10^{11}.

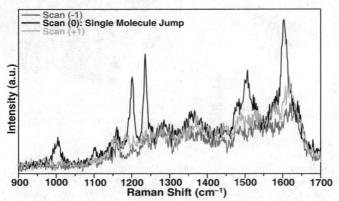

Figure 5. Three consecutive SERS scans with single molecule jump in Scan (0).

Table I shows the primary markers that are indicative of the transition from one conformation to another. These are: protonation markers (δCH, υC7=C8); trans/cis isomerization marker ($\upsilon\upsilon$C8-C9); and, H-bonding marker (υC9=O2) [10, 20, 21].

Table I. Transition markers for the photocycle of PYP: observed Raman peaks (VSCF calculation, DFT calculation) [10, 20-23].

	pG	pR	pB
$\upsilon\upsilon$C8-C9	1058 (1087, 1052)	998 (990, 993)	1002 (1000, 1002)
υC9=O2	1633 (1618, 1643)	1666 (1657,1642)	1653 (1674, 1660)
δCH	1163 (1173, 1146)	1165 (1147, 1157)	1174(1174, 1163)
υC7=C8	1558 (1608, 1547)	1556 (1555, 1537)	1576 (1581, 1502)
			1599 (1666, 1582)

The SERS spectra of single PYP molecules shown in Figure 6 and Figure 7 report that, the molecule was at the pG and pB intermediate states successively during the signal integration. In Figure 6, the peaks at 1152 and 1486 cm^{-1} are characteristic of the pG state and indicate that the pCA is ionized (deprotonated) [11, 21], while the peaks at 1172 and 1515 cm^{-1} are characteristic of the pB state and indicate that the pCA is protonated [11, 20]. Key markers that are indicative of the structural features of the pCA (Table I) were used to structurally interpret the single molecule SERS spectra of PYP. The data suggest that the PYP molecule was captured in a SERS hot spot while undergoing a transition between the pG and pB states of PYP. In a significant number of single molecule SERS spectra, peaks that are characteristic of the pB state were detected. Since these spectra were recorded in the dark and using a 514 nm laser line for the

Raman scattering, the accumulation of the pB photo-intermediate was not anticipated. These results suggest that, the 514 nm radiation is able to initiate the PYP photocycle at the surface-enhanced sites of the SERS substrate. The data also indicate that structural transitions (during the PYP photocycle) of a single PYP molecule can be monitored once it enters a hot site, although the molecule typically resides in the hot site for less than 1 s.

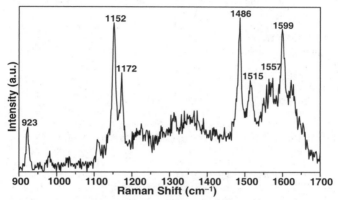

Figure 6. SERS signal, integrated for 0.25 s, capturing its protonation at the phenolic oxygen (from 1486 to 1515 cm^{-1}, from 1557 to 1599 cm^{-1}, and from 1152 to 1172 cm^{-1}) for a single PYP molecule [20-23].

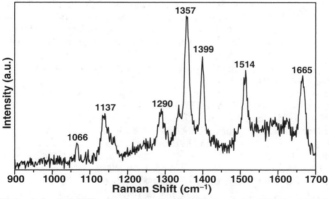

Figure 7. SERS signal of a single PYP molecule (integrated for 0.25 s) capturing: 1) the breaking of the hydrogen bond between the carbonyl oxygen and amide group of Cys69 (υC9=O2 mode shifting from 1626 to 1665 cm^{-1}); 2) protonation at phenolic oxygen (υCC peak shifts to 1515 cm^{-1}); 3) *trans*-isomerized state of pCA (corresponding to receiver state of PYP) (υC8–C9 peak at 1066 cm^{-1}); 4) strong sharp peak at 1357 cm^{-1}, attributed to υC9–O1 in the receiver state[20-23].

CONCLUSIONS

Single-molecule SERS of PYP was demonstrated using Ag nanoparticles reduced on Ge films. Single molecule spectra reveal various structural conformations and transitions of the PYP chromophore. These include cis/trans (isomerization), protonation/deprotonation, and H-bond breaking/re-establishment that typically occur during the photocycle of the PYP. These results will allow future analysis of the photocycle in PYP using single-molecule SERS studies with high structural sensitivity.

ACKNOWLEDGMENTS

The authors greatly appreciate funding of this work by Oklahoma State Regents for Higher Education.

REFERENCES

1. T. E. Meyer, *Biochim. Biophys. Acta* **806**, 175 (1985).
2. W. W. Sprenger, W. D. Hoff, J. P. Armitage, and K. J. Hellingwerf, *J. Bact.* **175**, 3096 (1993).
3. W. D. Hoff, P. Dux, K. Hard, B. Devreese, I. M. Nugterenroodzant, W. Crielaard, R. Boelens, R. Kaptein, J. Vanbeeumen, and K. J. Hellingwerf, *Biochemistry* **33**, 13959 (1994).
4. M. Baca, G. E. O. Borgstahl, M. Boissinot, P. M. Burke, D. R. Williams, K. A. Slater, and E. D. Getzoff, *Biochemistry* **33**, 14369 (1994).
5. T. E. Meyer, E. Yakali, M. A. Cusanovich, and G. Tollin, *Biochemistry* **26**, 418 (1987).
6. W. D. Hoff, I. H. M. Vanstokkum, H. J. Vanramesdonk, M. E. Vanbrederode, A. M. Brouwer, J. C. Fitch, T. E. Meyer, R. Vangrondelle, and K. J. Hellingwerf, *Biophys. J.* **67**, 1691 (1994).
7. M. Kim, R. A. Mathies, W. D. Hoff, and K. J. Hellingwerf, *Biochemistry* **34**, 12669 (1995).
8. A. H. Xie, W. D. Hoff, A. R. Kroon, and K. J. Hellingwerf, *Biochemistry* **35**, 14671 (1996).
9. G. E. O. Borgstahl, D. R. Williams, and E. D. Getzoff, *Biochemistry* **34**, 6278 (1995).
10. M. Unno, M. Kumauchi, J. Sasaki, F. Tokunaga, and S. Yamauchi, *Biochemistry* **41**, 5668 (2002).
11. A. H. Xie, L. Kelemen, J. Hendriks, B. J. White, K. J. Hellingwerf, and W. D. Hoff, *Biochemistry* **40**, 1510 (2001).
12. B. C. Lee, A. Pandit, P. A. Croonquist, and W. D. Hoff, *PNAS USA* **98**, 9062 (2001).
13. S. M. Nie and S. R. Emery, *Science* **275**, 1102 (1997).
14. K. J. Hellingwerf, J. Hendriks, and T. Gensch, *J. Phys. Chem. A* **107**, 1082 (2003).
15. M. A. Cusanovich and T. E. Meyer, *Biochemistry* **42**, 4759 (2003).
16. H. X. Xu, E. J. Bjerneld, M. Kall, and L. Borjesson, *Phys. Rev. Lett.* **83**, 4357 (1999).
17. C. E. Talley, J. B. Jackson, C. Oubre, N. K. Grady, C. W. Hollars, S. M. Lane, T. R. Huser, P. Nordlander, and N. J. Halas, *Nano Letters* **5**, 1569 (2005).
18. A. K. Kalkan and S. J. Fonash, *App. Phys. Lett.* **89**, (2006).
19. A. K. Kalkan and S. J. Fonash, *J. Phys. Chem. B* **109**, 20779 (2005).
20. M. Unno, M. Kumauchi, F. Tokunaga, and S. Yamauchi, *J. Phys. Chem. B* **111**, 2719 (2007).
21. M. Unno, M. Kumauchi, J. Sasaki, F. Tokunaga, and S. Yamauchi, *J. Phys. Chem. B* **107**, 2837 (2003).
22. D. H. Pan, A. Philip, W. D. Hoff, and R. A. Mathies, *Biophys. J.* **86**, 2374 (2004).
23. A. A. Adesokan, D. H. Pan, E. Fredj, R. A. Mathies, and R. B. Gerber, *JACS* **129**, 4584 (2007).

Mater. Res. Soc. Symp. Proc. Vol. 1077 © 2008 Materials Research Society   1077-L11-06

A Novel Near-Field Raman and White Light Imaging System for Nano Photonic and Plasmonic Studies

Ze Xiang Shen, J. Kasim, Y. M. You, and C. L. Du

Physics and Applied Physics, Nanyang Technological University, 1 Nanyang Walk, Blk 5 Level 3, Singapore, 637371, Singapore

ABSTRACT

We show the approaches in achieving high resolution Raman and white light imaging. In Raman imaging, a dielectric microsphere is trapped by the incoming laser, which was focused onto the sample by the microsphere. The microsphere was also used to collect the scattered Raman signals. We show the capability of this method in imaging various types of samples, such as Si devices and gold nanopattern. This method is comparatively easier to perform, better repeatability, and stronger signal than the normal near-field Raman techniques. Besides the Raman imaging, we also show a far-field confocal white light reflection imaging system that can be used for the fast imaging and characterization of nanostructures. This system uses a xenon (Xe) lamp as the incident light source and tunable aperture to enhance the spatial resolution. It has a spatial resolution of around 370 nm at a wavelength of 590 nm. With our system, we can clearly resolve images of 300 nm nanoparticles arranged in 2D honeycomb arrays with a period of 500 nm. Localized surface plasmons (LSPs) of isolated single and dimer gold nanospheres were also studied and the resonance energy difference between their LSPs was extracted.

INTRODUCTION

Raman spectroscopy measures molecular vibrations, which are determined by the structure and chemical bonding as well as the masses of the constituent atoms/ions. Raman spectra are unique in chemical and structural identifications. Conventional micro-Raman spectroscopy has a spatial resolution of about 500 nm, governed by the diffraction limit. To extend Raman imaging to the study of nano-materials, extensive efforts have been made to reduce the laser spot size below the diffraction limit by using scanning near-field optical microscopy (SNOM), which can be broadly divided into two approaches, laser delivered through an aperture [1–3] and tip-enhanced (apertureless) [4-6] near-field techniques.

Here we report a new approach, which we believe is a disruptive approach to near-field Raman microscopy. In this method, the laser is focused to a spot smaller than diffraction limit by a dielectric microsphere. Besides being used as the excitation source for Raman spectroscopy, the incident laser beam is also used to hold the microsphere just above the sample surface, through the well-known optical tweezers mechanism [7,8]. Simulation studies on optical nanojet based on plane wave incident light have shown that sub-diffraction limited focusing can be achieved when the diameter of the dielectric microsphere is comparable to the wavelength of laser [9,10]. Optical tweezers controlled 10 μm solid immersion lens (SIL) was used by A. L. Birkbeck *et al.* to perform optical microscopy on chrome grating [11]. Here we show the capability of trapping a dielectric microsphere to achieve high-resolution Raman imaging. This technique has many advantages over the previous near-field techniques. The Raman signal collected with microsphere using our technique is always much stronger than that without microsphere, by 2-7 times depending on the diameter of microsphere used [12]. This is a critical advantage over the aperture near-field technique. As the laser light is focused on the sample through the microsphere, there is no far-field signal in our setup, which has been one of the limitations in TERS. There is also no requirement to use a metal or metal-coated probe, e.g. metal-coated AFM tip, to perform the experiment. The strong near-field Raman signal and the simplicity in carrying out the experiment make this technique attractive, easy and fast. The reproducibility is also excellent, close to the 100% level. We also show that this technique can also work on different types of samples. This shows that this technique is versatile, easy to be implemented and reliable, making it extremely useful as nanocharacterization tool.

There is also growing interest in the study of the optical properties of metal nanostructures as the resonant excitation of coherent electron oscillations (commonly known as localized surface plasmons, or LSPs) differ greatly from that of the bulk material [13-15]. The LSP properties have rendered metal nanostructures useful for many applications, such as plasmonic crystals, surface-enhanced Raman scattering (SERS) and biosensing [16-21]. The size, shape and arrangement of metal nanostructures greatly affect the resonance energy of LSPs [22-24].

The optical response of metal nanostructures is strongly dependent on their corresponding structures. This dependence permits a more detailed understanding of the optical properties of the nanostructures. Accordingly, detecting and scanning metal nanostructures with a high spatial resolution are needed, so as to provide spatial-resolved information of the nanostructures. Optical scanning techniques offer complementary information to other scanning techniques, such as scanning electron microscopy (SEM) and atomic force microscopy (AFM). It is also non-invasive to samples, and is not limited to conductive materials. Among such techniques, scanning near-field optical microscopy (SNOM) provides simultaneous topographical and optical information of the sample with high resolution, and has been widely used. However, collecting an image by SNOM is very time-consuming and relies heavily on the equipment as well as the skill of the operator [25]. It is also not suitable for spectroscopic measurements due to the weak signals. Furthermore, it usually only allows for single wavelength scanning. On the other hand, far-field techniques are simpler and much more accessible. They have been successfully used to study LSPs of gold nanoparticle arrays [26,27]. Far-field scanning using either laser or white light excitation has a rather long history of development. Compared to laser scanning techniques, white light scanning is a simple and low cost method [28,29]. White light also offers multiple-wavelength advantage and can be used to perform absorption and reflection spectra. The spatial resolution of white light scanning can be greatly increased using the confocal technique. The highest spatial resolution for normal confocal white light (not including that from a super continuum light source [30]) scanning optical microscope has been improved from 1.5 μm to about 800 nm [31]. However, improvement in the spatial resolution is still much desired, for the study of small-scale materials.

We developed a new confocal white light reflection imaging system by combining a confocal white-light scanning microscope with a spectrometer. Its spatial resolution is about 370 nm, which is less than half of the previously reported spot size of ~800 nm for white light scanning; and is even smaller than that of some laser scanning techniques [32,33]. Although this spatial resolution is still worse than that of SNOM, our system is low cost, easy to use, and it also generates much stronger signals. Most importantly, it can perform simultaneous imaging and acquisition of white light reflection/absorption spectra, which offers vital information that relates the optical properties to their corresponding nanostructures.

Thus far, single nanoparticle or nanowire imaging and spectroscopy experiments have mainly focused on laser or electron excitation [34,35]. For white light excitation, the experiments are normally operated under dark field illumination [22] or use internal reflection microscopy [36]. In our work, we used a Xe lamp as the excitation source and bright field illumination. The spatial resolution of our system was determined using a scanning knife-edge method and checked using gold particle arrays fabricated by nanosphere lithography (NSL). Using this system, we also present the white light reflection images and spectra of gold nanospheres – both single and dimer, with 50 nm diameter; as well as those of silver nanowires with 100 nm diameter. The excitation of the LSPs of the spheres and nanowires could be inferred from the results.

EXPERIMENT

The near-field Raman microscopy setup with polystyrene microsphere (diameter = 3 μm) is based on the WITec CRM200 confocal Raman microscopy system (25 μm pinhole) with microscope objective. A schematic diagram of the experimental setups is shown in figure 1.

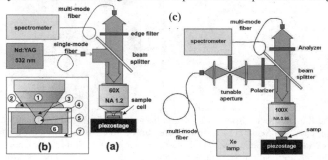

Figure 1. *(a)* Schematic diagram of the near-field Raman microscope with microsphere. *(b)* Detailed description of the sample cell: (1) water immersion lens (60X NA=1.2 WATER IMMERSION), (2) water, (3) focused laser, (4) cover glass, (5) polystyrene microsphere (3 mm), (6) sample, and (7) sample cell. *(c)* Schematic diagram of the confocal white light reflection imaging system.

For our high resolution Raman setup, a double-frequency Nd:YAG laser (532 nm) is used as the excitation laser, as shown in Figure 1a. The linearly polarized laser that is used to excite the Raman signal is also used to optically trap the microsphere. The laser beam is incident on the sample through the microsphere. Sample is placed in a sample cell with diluted polystyrene microspheres in water. The sample cell is put on a translation stage, which can be moved coarsely along *x-* and *y-axes*. It also can be finely moved with a piezostage. The Raman scattered light was directed to a grating (1800 grooves/mm or 600 grooves/mm) and detected using a TE-cooled charge-coupled-device (CCD). Almost similar setup can be used for white light imaging, as shown in Figure 1c. A tunable aperture with a minimum diameter of 200 μm was introduced in the incident light path to enhance the spatial resolution. Light from a Xe lamp was polarized after passing through a polarizer. The incident light was focused onto the sample through a beam splitter and an OLYMPUS microscope objective lens (100X, NA=0.95). The reflected light from the sample was collected by the same lens and directed to a spectrometer through a fiber. Fibers with various core diameters of 100, 50 and 25 μm were employed in this work. The collection fiber also works as a pinhole, which is confocal with the illuminated spot on the sample and blocks light reflected from objects that are not at the focal point. The reflected light was directed to a 150 grooves/mm grating and detected by a CCD.

RESULTS AND DISCUSSION
1. High resolution Raman imaging

In this paper, we focus on the study of PMOS transistors with SiGe source drain stressors and poly-Si gate. The patterned wafers used in this study were prepared using 65 nm device technology. After spacer formation and Si recess etch, the wafers were cleaned and the epitaxial SiGe growth was performed on a commercially available low-pressure chemical vapor deposition (LPCVD) system.

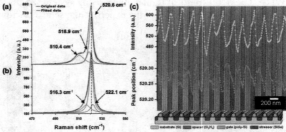

Figure 2. The Raman spectra from *(a)* SiGe and *(b)* poly-Si lines with fitted peaks using Lorentzian function. *(c)* Scanning Electron micrograph with cross-section view diagram of periodic poly-Si lines and SiGe stressors. Line-scan of Raman Si-Si intensity from SiGe is shown in yellow color, and the Si-Si peak position from the bulk Si is in purple color. The line scans show excellent correspondence with the structure.

Fig. 2(a) and 2(b) show the Raman spectra recorded in the SiGe line region and on top of the poly-Si line, respectively. Each spectrum was fitted with three Lorentzian peaks. In Fig. 2(a), the Raman peaks from SiGe line correspond to Si-Si phonon vibrations from the SiGe (510.4 cm^{-1}), tensile-strained Si just below the SiGe (518.9 cm^{-1}), and the Si substrate below (520.6 cm^{-1}), respectively. Similarly, in Fig. 2(b), the Raman peaks correspond to Si-Si phonon vibrations of poly-Si (516.3 cm^{-1}) and bulk Si below (520.6 cm^{-1}), and another one is from compressively strained Si in the channel region (522.1 cm^{-1}). Fig. 2(c) shows the SEM image of the device sample together with the detailed illustration diagram of the device structure. From Fig. 2(c) we can see the line profile of the integrated intensity of Si-Si phonon vibrations from the SiGe (yellow color), and the Si-Si peak position from the bulk Si (purple color). The results show excellent correspondence with the device structure with good S/N ratio.

We have also performed Raman mapping on gold nanopatterns. Gold nanopatterns were fabricated on silicon substrate. The size of the gold nanopatterns is ~100 nm.

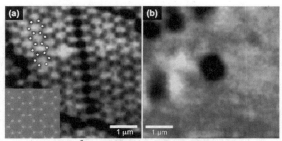

Figure 3. *(a)* and *(b)* are 5.0x5.0 μm² near-field and confocal Raman images of the gold nanopatterns, respectively, generated using Si Raman peak intensity from the substrate. The white dots illustrate the gold nanopatterns on the substrate. Inset shows the SEM image of the gold nanopatterns with the size of the nanopatterns is ~100 nm. The scale bar denotes one micron.

Fig. 3(a) and 3(b) show the 5.0x5.0 μm² (100x100 points) near-field and confocal Raman images of gold nanopatterns, respectively. Fig. 3(a) shows the Raman image from the Si-Si peak intensity. Darker regions correspond to the gold nanopatterns. It can be clearly seen that the gold nanopatterns with the size of ~100 nm can be clearly resolved using our technique. The near-field Raman image corresponds well to the electron micrograph as shown in the inset of Fig. 3(a). The confocal Raman image of the gold nanopatterns shows no details of the patterns, as shown in Fig. 3(b). This proves the capability of our technique to various samples. Besides this, the spatial resolution can be improved further by reducing the Brownian motion of the microsphere in the solution, e.g. in sugar solution.

2. White light reflection imaging

To illustrate the white light imaging capability of our system, reflection images of gold nanoparticle arrays are shown in Fig. 4(a-d). The gold particle arrays were fabricated by NSL [37] with polystyrene (PS) microspheres as masks. The microspheres have diameters of 0.5 μm and 1 μm, as shown in Fig. 4(a-b) and Fig. 4(c-d) respectively. The thickness of the nano arrays is about 50 nm. In Fig. 4(a), the bright round spots come from the silicon substrate while the dark areas come from the gold particles since the silicon surface reflects more light than the gold particles at the present wavelength range of 480-520 nm. The reason why this wavelength range is chosen will be discussed later. The 0.5 μm periodicity of the gold particle honeycomb-like arrays can be clearly recognized, demonstrating the high resolution of our system. The turquoise dots in the image are a guide for the eye to locate the particles. In contrast to the case of Fig. 4(a), the bright, hexagonal ring-like pattern in Fig. 4(b) corresponds to the gold particles while the dark areas correspond to the cover glass instead because the cover glass substrate reflects less light than the gold particles. The size of the gold particles and the center-to-center distance between two nearest gold particles are about 150 nm and 250 nm respectively, which are barely distinguishable in our results. Defects in the samples as indicated by the turquoise curves in Fig. 4(a and b) can be clearly seen in the images, demonstrating that the present imaging method can also be used to test the sample quality as shown by D. O. Anjeanette *et al.* [38], but with much better spatial resolution here.

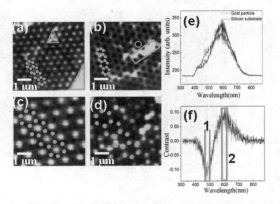

Figure 4. The confocal white light reflection images (a-d) and spectra (e, f) of gold particle arrays. (a) and (b), images of gold arrays at λ = 480-520 nm wavelength region on silicon and cover glass, respectively, which were fabricated by using 0.5 μm diameter PS as the lithographic mask; (c) and (d), images of gold arrays fabricated by using 1 μm diameter PS as the lithographic mask on silicon; (e) and (f), the white light reflection spectra and the reflection contrast spectra, respectively, obtained from (c) and (d). Wavelength regions for images (c) and (d) are 460-500 nm and 580-620 nm, respectively, labeled as green rectangles 1 and 2 in (f). Turquoise dots in (a) to (d) are guides to label where the gold particles are. The scale bar denotes one micron.

The present system is capable of multi-wavelength imaging. The images at 460-500 nm and 580-620 nm are shown in Figs. 4(c) and (d) respectively, for gold nanoparticle patterns fabricated on silicon using 1 μm diameter PS spheres. The images were taken at the wavelength of maximum positive contrast near 600 nm and maximum negative contrast around 480 nm (Fig. 4(f)). All the images in this paper were selected by this way. Both the dark areas in Fig. 4(c) and the bright dots in Fig. 4(d) originate from the gold particles, demonstrating the wavelength-dependent nature of reflectivity for both gold particles and silicon substrate. The gold particle size and center-to-center distance between two nearest nanoparticles were about 300 nm and 500 nm respectively, as measured from SEM images (not shown). From Figs. 4(c) and (d), the six gold nanoparticles in one hexagon (labeled by the turquoise dots in the figures) can be clearly recognized. The white light reflection spectra of the substrate and the gold particles (Fig. 4(e)) can be easily extracted from the images and the contrast spectrum (Fig. 4(f)) as defined below in Eq. (1) below can be derived:

$$\text{Constrast} = (I_{\text{sample}} - I_{\text{substrate}}) : I_{\text{substrate}} \tag{1}$$

where I_{sample} and $I_{\text{substrate}}$ refer to the white light reflection intensity of the sample and the substrate respectively. Fig. 4(f) exhibits a dip and a peak at about 480 nm 600 nm, respectively, which manifests the maximum reflection contrast at these two wavelengths.

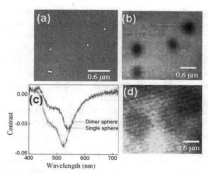

Figure 5. SEM image (a) and confcoal white light reflection images at 510-550 nm wavelength region of gold spheres on 200 nm silicon dioxide films with collection fiber core diameter 25 μm (b) and 100 μm (d). (c), the contrast spectra for single and dimer spheres. Scale bar denotes 0.6 μm.

The above results readily demonstrate the potential of our system to effectively probe the contrast images and contrast spectra of nanostructures. More importantly, some effects of the surface plasmons on reflection can also be elucidated by studying the contrast spectra, which is discussed in detail in the following sections.

As shown in Fig. 5(a), the SEM image of the gold nanospheres contains three isolated single spheres and one dimer where the two spheres are almost in contact with each other. Fig. 5(b) gives the confocal white light reflection images for the corresponding spheres in the wavelength region of 510-550 nm. Four dark spots represent the images of the gold spheres, which correspond well to their SEM image. The size of the dark spots is about 400 nm, approximating to the spatial resolution of the system. Applying Eq. (1) to the single spheres and the dimer, the contrast spectra which exhibit two dips (labeled 1 and 2 respectively) were obtained as shown in Fig. 5(c). For isolated single spheres, the wavelength of 525 nm at dip 2 coincides with that of the surface plasmon of 50 nm gold spheres [39]. However, the same dip shows a red shift to 540 nm for the dimer. This is a result of the coupling effect between the two spheres in the dimer, which is in agreement with literature [40]. The position of the weak dip 1 for the single sphere almost overlaps with that for the dimer located at about 480 nm. We believe this dip originates from the multi-polar surface plasmon excitation of the gold spheres. Surface plasmon excitation at about 480 nm has been reported for gold nanospheres with diameters close to 40 nm, but was ascribed to false spectral lines that arose from using a 488 nm argon laser [41]. The excitation of LSP leads to the enhancement in the absorption of the particles, which consequently has the effect of reducing their reflection intensity. The system cannot resolve the spheres when the 100 μm core diameter collection fiber is used as demonstrated by the images in Fig. 5(d). This again confirms that we can achieve a higher spatial resolution by using a 25 μm core diameter collection fiber.

CONCLUSIONS

In conclusion, we report a new approach in performing high-resolution near-field Raman imaging. High-resolution Raman image of PMOS transistors with SiGe source drain stressors was obtained by scanning a 3 μm diameter polystyrene microsphere using optical tweezers mechanism. The microsphere is used to focus the excitation laser, and also to collect the scattered Raman signal. The major advantages of this technique are non-destructive, high reproducibility (almost 100%), fast (strong signal), no far-field background, and easy to use compared to other near-field Raman techniques, e.g aperture and apertureless methods. We also showed the capability of this technique in studying the strain on sub-100 nm semiconductor device, in which Si channel is compressively strained by SiGe stressors. Besides on the device sample, high-resolution Raman imaging was also performed on gold nanopatterns on Si substrates. The simplicity and reproducibility of this approach to achieve high resolution Raman imaging will make it attractive to be used for large-scale applications in nano-science and nano-technology. For the white light imaging system, we have developed a simple, non-invasive and easy-to-use system, which can perform simultaneous imaging and acquisition of reflection spectra of samples with a spatial resolution about 370 nm. With it, even isolated single and dimer gold nanospheres with a diameter of 50 nm can be imaged, in which multi-polar LSP excitation was revealed apart from dipolar LSP. Compared to the resonance energy for the single sphere, the resonance energy for the dimer is red-shifted due to the coupling between the two spheres in the dimer. As the white light imaging can be performed at different wavelengths, we expect interesting applications such as biomaterial mapping and plasmonic studies.

ACKNOWLEDGMENTS

The authors would like to thank Nanyang Technological University for the financial support.

REFERENCES

1. D. W. Pohl, W. Denk, and M. Lanz, *Appl. Phys. Lett.* **44**, 651 - 653 (1984).
2. B. Hecht, H. Heinzelmann, and D. W. Pohl, *Ultramicroscopy* **57**, 228 - 234 (1995).
3. J. Kim, J. H. Kim, K. B. Song, S. Q. Lee, E. K. Kim, S. E. Choi, Y. Lee, and K. H. Park, *J. Microsc.* **209**, 236 - 239 (2003).
4. F. Zenhausern, Y. Martin, and H. K. Wickramasinghe, *Science* **269**, 1083 - 1085 (1995).
5. D. H. Pan, N. Klymyshyn, D. H. Hu, and H. P. Lu, *Appl. Phys. Lett.* **88**, 093121 (2006).
6. H. G. Frey, C. Bolwien, A. Brandenburg, R. Ros, and D. Anselmetti, *Nanotechnology* **17**, 3105 - 3110 (2006).
7. A. Ashkin, *Science* **210**, 1081 - 1088 (1980).
8. A. Ashkin, *Proc. Natl. Acad. Sci. USA* **94**, 4853 - 4860 (1997).
9. X. Li, Z. G. Chen, A. Taflove, and V. Backman, *Opt. Express* **13**, 526 - 533 (2005).
10. S. Lecler, Y. Takakura, and P. Meyrueis, *Opt. Lett.* **30**, 2641 - 2643 (2005).
11. A. L. Birkbeck, S. Zlatanovic, S. C. Esener, and M. Ozkan, *Opt. Lett.* **30**, 2712 - 2714 (2005).
12. K. J. Yi, H. Wang, Y. F. Lu, and Z. Y. Yang, *J. Appl. Phys.* **101**, 063528 (2007).
13. S. R. Emery, W. E. Haskins, S. M. Nie, *J. Am. Chem. Soc.* **120**, 8009-8010 (1998).
14. M. J. Feldstein, C. D. Keating, Y. H. Liau, M. J. Natan, N. F. Scherer, *J. Am. Chem. Soc.* **119**, 6638-6647 (1997).
15. S. Kawata, *Near-field Optics and Surface Plasmon Polaritons,* S. Kawata, M. Ohtsu, M. irie, Eds. (Springer, 2001).

16. N. Grigorenko, A. K. Geim, H. F. Gleeson, Y. Zhang, A. A. Firsov, I. Y. Khrushchev, J. Petrovic, *Nature* 438, 335-338 (2005).
17. A. N. Grigorenko, H. F. Gleeson, Y. Zhang, N. W. Roberts, A. R. Sidorov, A. A. Panteleev, *App. Phys. Lett.* **88**, 124103 (2006).
18. S. C. Kitson, W. L. Barnes, J. R. Sambles, *Phys. Rev. Lett.* **77**, 2670-2673 (1996).
19. A. J. Haes and R. P. Van Duyne, *J. Am. Chem. Soc.* **124**, 10596-10604 (2002).
20. A. G. Brolo, E. Arctander, R. Gordon, B. Leathem, K. L. Kavanagh, *Nano Lett.* **4**, 2015-2018 (2004).
21. J. Grand, M. Lamy de la Chapelle, J.-L. Bijeon, P.-M. Adam, A. Vial, P. Royer, *Phys. Rev. B* **72**, 033407 (2005).
22. L. J. Sherry, R. Jin, C. A. Mirkin, G. C. Schatz, R. P. Van Duyne, *Nano Lett.* **6**, 2060-2065 (2006).
23. G. H. Chan, J. Zhao, E. M. Hicks, G. C. Schatz, R. P. V. Duyne, *Nano Lett.* **7**, 1947-1952 (2007).
24. C. Noguez, *J. Phys. Chem. C* **111**, 3806-3819 (2007).
25. I. Notingher and A. Elfick, *J. Phys. Chem. B* **109**, 15699-15706 (2005).
26. G. Laurent, N. Félidj, S. Lau Truong, J. Aubard, G. Lévi, J. R. Krenn, A. Hohenau, A. Leitner, F. R. Aussenegg, *Nano Lett.* **5**, 253-258 (2005).
27. G. Laurent, N. Félidj, J. Grand, J. Aubard, G. Lévi, A. Hohenau, F. R. Aussenegg, J. R. Krenn, *Phys. Rev. B* **73**, 245417 (2006).
28. E. Pecheva, P. Montgomery, D. Montaner, L. Pramatarova, *Langmuir* **23**, 3912-3918 (2007).
29. Z. H. Ni, H. M. Wang, J. Kasim, H. M. Fan, T. Yu, Y. H. Wu, Y. P. Feng, Z. X. Shen, *Nano Lett.* **7**, 2758-2763 (2007).
30. K. Lindfors, T. Kalkbrenner, P. Stoller, V. Sandoghdar, *Phys. Rev. Lett.* **93**, 037401 (2004).
31. Y. Youk and D. Y. Kim, *Opt. Commun.* **262**, 206-210 (2006).
32. C. Rembe and A. Dräbenstedtb, *Rev. Sci. Instrum.* **77**, 083702 (2006).
33. L. Gütay and G.H. Bauer, *Thin Solid Films* **515**, 6212-6216 (2007).
34. A. Cvitkovic, N. Ocelic, R. Hillenbrand, *Nano Lett.* **7**, 3177-3181 (2007).
35. M. Bosman, V. J. Keast, M. Watanabe, A. I. Maaroof, M. B. Cortie, *Nanotech.* **18**, 165505 (2007).
36. C. Sönnichsen, S. Geier, N. E. Hecker, G. von Plessen, J. Feldmann, H. Ditlbacher, B. Lamprecht, J. R. Krenn, F. R. Aussenegg, V. Z-H. Chan, J. P. Spatz, M. Möller, *Appl. Phys. Lett.* **77**, 2949-2951 (2000).
37. T. Jensen, M. Duval, K. Kelly, A. Lazarides, G. Schatz, R. Van Duyne, *J. Phys. Chem. B* **103**, 9846-9853 (1999).
38. A. D. Ormonde, E. C. M. Hicks, J. Castillo, R. P. V. Duyne, *Langmuir* **20**, 6927-6931 (2004).
39. M. A. V. Dijk , M. Lippitz , M. Orrit , *Acc. Chem. Res.* **38**, 594-601 (2005).
40. A. Moores and F. Goettmann, *New J. Chem.* **30**, 1121-1132 (2006).
41. S. Benrezzak, P. M. Adam, J. L. Bijeon, P. Royer, *Surf. Sci.* **491**, 195-207 (2001).